Investing in MUTUAL FUNDS *Using* FUZZY LOGIC

by

KURT PERAY

S^t^L

St. Lucie Press

Boca Raton London New York Washington, D.C.

Investing in MUTUAL FUNDS *Using* FUZZY LOGIC

Library of Congress Cataloging-in-Publication Data

Peray, Kurt.
Investing in mutual funds using fuzzy logic / Kurt Peray.
p. cm.
ISBN 1-57444-264-3 (alk. paper)
1. Mutual funds. 2. Fuzzy logic. I. Title.
HG4530.P42 1999
332.63'27--dc21 99-12298
CIP

Direct all inquiries to CRC Press LLC, 2000 Corporate Blvd., N.W., Boca Raton, Florida 33431.

St. Lucie Press is an imprint of CRC Press LLC

International Standard Book Number 1-57444-264-3
Library of Congress Card Number 99-12298
Printed in the United States of America 1 2 3 4 5 6 7 8 9 0
Printed on acid-free paper

Table of Contents

Section II: Using Fuzzy Logic

Foreword

by Chemical Publishing Co., Inc., New York

This book sets out a new investment approach developed by Mr. Kurt Peray, based on the Fuzzy Logic concept that he successfully applied to the process control of cement kilns.*

The book is aimed at the individual investor who wants to invest in financial instruments that will provide a return for growth. While making her/his own independent and educated decisions, the investor will be able to hedge her/his portfolio from the volatile forces in the market, and will offset the erosive impact of inflation and taxation. Mr. Peray's detailed analysis of the new approaches and techniques for risk and portfolio asset allocation will enable an investor to make decisions as to when to buy, hold, or sell financial instruments.

In Mr. Peray's words: "There are no promises for spectacular and quick profits, neither does this strategy engage in futile attempts to predict the future. It shuns any kind of speculative investments and places the emphasis on preservation of capital."

In this electronic age, investors have quick access to very important information relevant to the investment decision process. The guidelines and formulas that serve as foundations to the Fuzzy Logic approach will enable the investor to build customized computer programs. Although computers are not a requirement to implement the Fuzzy Logic technique, they are a valuable tool to speed up the data collection.

---

* Peray, K. E., *The Rotary Cement Kiln,* Chemical Publishing Co., Inc., New York, 1970.

Preface

Are the financial markets manipulated by wealthy individuals, arbitragers, analysts, and brokerage houses? Does an individual investor of modest means, have any chance to make an honest profit in a stock market environment that seems to be controlled by large institutions using computerized program trading and slick advertisement techniques? Could it be that options, futures, and other derivative trading, instead of smoothing market volatility, could be the main reasons for the frequent large changes in the market indexes we have noticed lately? Doesn't the irrational behavior of the markets suggest that stock markets have become irrational and a giant casino? Does it make sense when a company reports a 24% increase in earnings and its stock tumbles the next day? Or when another company reports a loss in the millions for the quarter and its stock increases more than 1% immediately after the loss is made public, simply because analysts have forecast much larger profits in the first case and greater losses in the second.

Where is the rationale when: a) the head of one of the largest banks in the world loses billions because he allowed his bank to lend its name for questionable and risky speculative hedge fund investments; b) this caused his bank, its employees, and its stockholders to lose millions because of his greed for quick profits and disregard of the risks associated with such speculative investments; c) thousands of bank employees lose their jobs because of this and, finally, d) he is rewarded several millions in form of golden parachutes because he "voluntarily" resigned his position after the fiasco? Where is the rationale? Where is the justification in giving him such a going-away present? It's weird — doesn't it look like a zoo out there in the investment and banking world?

These are some legitimate questions investors are entitled to ask. Evidence seems to support the contention that program and derivative trading has added volatility to the markets and seems to benefit primarily large institutions and the wealthy at the expense of the small investor.

The image the general public has of the financial markets is generally negative during severe market corrections, and positive, bordering on euphoria, during stampeding bull markets. Such attitudes prevailed during and after the market crashes of 1929, 1987, and 1998. We can be sure these attitudes will again prevail in the future under similar circumstances.

Contrary to their responsibilities, government regulatory agencies will likely not alleviate such market excesses for they tend to react only after a major crisis has erupted. Decisive action and regulations are usually late in coming and normally will take place only when the public exerts political pressure.

Most small investors rely on broker or banker advice for investment recommendations and most believe the professional management of mutual funds will protect them from major financial losses in a crisis. This is a bad assumption as we will learn later in this book. Investors are constantly bombarded with all kinds of conflicting information and market predictions by the media, but no smooth talk nor advertising hype will ever prevent scores of investors from significant portfolio losses in a market crash that could happen again at any time. At any time, a market guru might predict a higher market, another, exactly the opposite. Sad to say that there will always be a sizable number of small investors who take this advice at face value and often make their investment decisions based on it. The fact is, nobody can predict the future.

Many a small investor has become so discouraged by this that they decide to stay on the side lines and let the experts play their own games. Government officials often enact regulations and rules that seem to benefit the banks or investment brokerage houses more than the general public. Government favoritism, ineptitude, and neglect to come to grips with financial problems have cost taxpayers dearly and pushed the economies in many countries to the brink of insolvency. One doesn't have to look too far back to the speculative real estate bubbles in Japan and Hong Kong, the debt crisis in Mexico or South Korea, and the S&L and hedge fund debacles in the U.S. Unquestionably, more will be added to this list in the future.

No wonder small investors are concerned about market volatility and what it could do to their hard earned savings. Some might even lose sleep over it. Surely there must be other things more important in life than having to continuously worry about one's own investments.

Many financial advisors advocate buying and holding for the long term and disregarding the severe ups and downs even if they last for one year or more. Looking back over the past ten years, the most dramatic bull market in history, this theory might have held true. One doesn't need to be a rocket scientist to make money in a raging bull market. Anyone can do it by just following the herd. But there is a flip side we must also consider and always keep in mind. Namely, bull markets don't last forever. There will be times, just like in the 1970s, when mutual fund investors will have to wait for several years to recoup their losses in a bear market. When that will materialize, nobody knows. It could happen next week, month, or in a few years into the future. So, the buy-and-hold concept is not the answer. Ignoring market swings and holding for the long term is also not a solution for investors advanced in age. The good times when an investor could use this passive approach toward investments seem to be over. Investments have to be monitored more closely than ever before and one must be prepared to take sudden and swift action whenever signs of speculative excesses in the market are present.

We want to be prudent investment mountain climbers, unlike the inexperienced tourists who hire a mountain guide to climb the Matterhorn but they only wear sneakers and shorts. They might reach the peak, all the while forgetting they have a long and dangerous way down ahead of them. It's always easier to climb a mountain than to come down.

The opportunities the capitalistic system provides each one of us to acquire more wealth and financial freedom carries a price. But, despite all the weaknesses in the system, one can still reap the rewards for being a frugal saver and clinging to a prudent investment program. While nobody can predict the future, adherence to some basic principles of investing will allow even the small investor to invest successfully for his future financial needs.

What is one to do? Forget about investments and put the money into a "secure" savings account in a bank or store it in a shoe box? We certainly hope not; this would be tantamount to an ostrich sticking his head into the sand. Get into debt in the hope that inflation will take care of the repayment in the future? Live on credit and buy all the things we crave before we actually earned the money for it? We don't want to make the same mistakes some financial high fliers made in the 1980s. As private investors we certainly cannot afford to make highly speculative investment decisions. It's our investment money and we cannot afford to place it blindly in someone else's hands.

Amazingly, most people will generally not give their cars, boats, or houses for use to strangers but many are perfectly willing to hand over their entire life savings to a complete stranger that names himself a "financial advisor" or "investment manager," sometimes even over the phone without ever having checked the legitimacy and credentials of such telemarketing operators. Let us never forget that these operators are out first and foremost to generate commissions for themselves.

Whatever investment action we take could either hurt or benefit our financial well-being. We do not have the luxury of speculating with OPM, or "other people's money." Unfortunately, as small investors, we are forced to participate in the security markets because most CDs and money market instruments don't provide us with the returns needed to offset the ravages of inflation and taxation.

This book describes a new investment philosophy that disregards the hype of the financial community and shows the small investor a way to make his own independent investment decisions. This is an investment strategy for financial security based on common sense and in many respects quite different from what is currently accepted practice in portfolio management. There are no promises for spectacular and quick profits, neither does this strategy engage in futile attempts to predict the future. My method teaches the reader to make his investment portfolio grow at a steady, realistic pace. It shuns any kind of speculative investments and places the emphasis on preservation of capital. The concept presented herein runs in many respects contrary to what is being taught today in universities and investment seminars. It definitely is not for the investor on the lookout for a quick 25% profit in a few month's time.

Stock brokers and investment advisors might argue the validity of our approach and certainly the financial high fliers will look at this concept as a dull exercise. Let them have their own beliefs; after all, we are all free to make up our own minds. Let them play games with someone else's money as long it is not our own.

The technique presented here is not new; the author has published the TOO HIGH-OK-TOO LOW method in 1970 which later become known as the Fuzzy Control concept.* There is nothing "fuzzy" about our approach as I will discuss in details in this book. Initially introduced for the purpose of process control, the author has applied the same principles for his investment decisions and they have proven to be useful and valid for control of his own

* Peray/Waddell, *The Rotary Cement Kiln*, Chemical Publishing Co., Inc., New York, 1970.

portfolio as well. The Fuzzy Control concept has served the author well for over 20 years and has produced consistent returns in each of these years. We never had a negative total return in our portfolio, neither in the year of the crash, 1987, nor in other years when the stock and/or bond markets were down.

The name Fuzzy Control was first coined by a European kiln manufacturer who used the TOO HIGH-OK-TOO LOW concept described in my earlier book to develop a successful computer program for the purpose of automatic kiln control. Fuzzy Control is nothing more than a simplified technique to put down on paper the kind of decision process (logic) an operator has to go through when two, three, or four variables interact with each other to produce a certain outcome. I know it sounds complicated, but it is not. Let me explain.

The Dow Jones Industrial Average, as most investors will know, moves in different directions as a result of perhaps millions of reasons and variables exerting an influence on this market index. There is no way anybody could ever measure all these variables and thus come up with a mathematical model that accurately and consistently predicts what the market will do next week, month, or year. However, as a hypothetical example, assume that movements in the Dow Jones Industrial Average are predominantly governed by the interaction of three variables, (Chapter 8 explains these variables in detail) namely the Price-to-Earnings Ratio (P/E), interest rates, and current state of the economy. Now, each of these three variables can be too high, too low, or okay and, thus this market index is exposed to a total of $3 \times 3 \times 3 = 27$ basic conditions, each of which would trigger a certain and different reaction in this index.

Many investors have learned that such correlations exist and are using their own skills and experiences to guide them in their investment decisions depending on the status of these variables. For example, an investor might conclude that, when the market's P/E ratio is too high, interest rates are rising and the economy is turning down, and it would be the time to sell equities, irrespective of the current market trend or what the market gurus are predicting. His mind is employing the logic: P/E high, interest rates higher, economy lower to make the decision to sell. In essence, he has been using what is referred to as Fuzzy Logic and this is the technique we will discuss in details in this book. It's a simplified method to solve problems that, at first glance, appear to be too complex and unsolvable.

Computers work with the binary system — it's either a 1 or a 0, buy or sell, yes or no, high or low. In other words, it is two-dimensional in scope. The Fuzzy Control concept however, also recognizes an OK condition where a variable is neither too high nor too low, and places as much value on this OK condition as on the other two. Thus, with Fuzzy Logic, a third dimension is added. Fuzzy Control is now being successfully employed for computer control of industrial processes and has become an accepted control technique worldwide in many industries. A Japanese manufacturer produces computer chips that use the same Fuzzy Logic concept. The author strongly believes that this concept has far-reaching benefits in those applications where human intelligence, experience, and logic are still needed and thus is capable of outweighing the most sophisticated computer programs.

The concept discussed herein is not the ultimate panacea, but from the author's experiences, represents a fresh approach so badly needed in the financial community. It eliminates the hype and mania spread around by some financial institutions and brokerage houses. Our technique allows the investor to make his own independent investment decisions free of outside influences.

We live in an age where investors make increased use of computers and have quick access to data important to the investment decision process. For this reason we have included guidelines and formulas for the readers to construct their own computer programs that serve as the foundation of the herein discussed Fuzzy Logic investment technique. Although a computer is not a requirement for this technique, it nevertheless helps to speed up data collection. The technique discussed herein describes not only when to buy, hold, or sell financial instruments, but also presents new approaches and techniques developed by the author for risk control and portfolio asset allocations.

Many savvy investors have become financially independent. So can the reader. But, it will take discipline, sacrifices, and a solid commitment to strive for and achieve this goal. There are no quick fixes for a secure financial future. It's a long way up. We must climb the mountain to success on our own, step by step. And, most important, we must have the stamina and staying power to stick to our savings and investment strategy for the long term.

This book can guide the reader toward financial security and make him comfortable in the world of finances. It shows ways to reduce risks, preserve the capital, and attain a rate of return that exceeds the rate of inflation and taxes. We discuss here a common sense approach to investing. Part I discusses assessing your financial situation, how much we'd like to retire with, and what funds can help achieve that goal. Part II shows how to apply this data and how our Fuzzy Logic concept can further help achieve your goals. There is one advantage we have when compared to brokers, professional portfolio managers, and bankers; namely, we are not under pressure to produce a given return in a given quarter or year. We don't have to outperform the market averages and we don't have to frequently turn over our portfolio to make it look good on the prospectus at the end of a quarter or fiscal year.

To be a successful investor one has to take the first step on his own; namely, one has to commit one's self to a disciplined program of savings and to bring one's financial house in order. Nobody can help us in this endeavor for we ourselves must make that initial commitment. In the absence of such a commitment, our dream of a secure financial future will remain just that — a dream.

About the Author

Kurt Peray, a Chemical Engineer and Marketing Manager with 38 years of professional experience, has recently retired and now lives with his wife in Florida.

Mr. Peray has developed an investment technique based on the Fuzzy Logic principles, which he has successfully applied for over 25 years in his capacity as an investment portfolio advisor for several private individuals.

Kurt Peray is the author of "Cement Manufacturer's Handbook" and "The Rotary Cement Kiln," published by Chemical Publishing Co., Inc., works highly regarded in the chemical engineering field. First published in 1981 and 1970, respectively, both titles continue to be in demand worldwide.

Acknowledgments

My thanks to my wife Sonja, my children Daniela and Marcel
for their patience and forbearance.

Also my thanks to Silvia Soto-Galicia who believed in this project
from the onset and to Marie Etzler, my editor at CRC Press
for her guidance and support from conception to completion of this book.

Kurt Peray
January 1999

I

PREPARING TO USE FUZZY LOGIC: GATHERING DATA

1 Overview of the Personal Financial Situation

One of the Texas billionaire Hunt brothers, questioned about the size of his wealth, has uttered the now famous answer: "You're not rich if you know how much money you have." I seriously question the wisdom of such an answer. If the government would ever institute a wealth tax, which is not as far fetched as it might sound, one can be sure these super rich individuals would know exactly how much they are worth, down to the last cent.

Nobody can afford not to know his personal net worth. The key word here is net. A "wealthy" individual might own ten skyscrapers, nine shopping centers, eight hotels valued 1.2 billion in total, but he might also be in debt to the tune of 1.3 billion. Just because someone owns a 120-foot yacht, travels in his own Lear Jet, and lives in a 30-room mansion doesn't necessarily mean the guy can set claim to a high net worth. Neither can such an individual be regarded as being filthy rich. The fact is that many of these so-called fortunes are mortgaged and leveraged. Some are not much different than a middle class family that owns five credit cards with each account run up to the debt limit. The only difference might be that the "super rich" guy has a few more zeros attached to his balance sheet and that bankers look at him in awe.

1.1 Determination of Net Worth

Net worth is the amount left over after all the liabilities are subtracted from the assets an individual holds. To determine this amount, one can make use of a work-sheet as shown in Table 1.1. It is not uncommon for people to be

Table 1.1 Determination of Net Worth

Assets		*Liabilities*	
A) Liquid assets			
Checking account		Current bills to pay	
Cash on hand		Credit card balances	
Money market accounts		Installment loans	
Savings accounts		Personal loans	
Others		Business loans	
		Annuity contracts	
B) Frozen assets		Life insurance	
Equities		Home mortgage	
Bonds		Alimony, child support	
CDs and gov't. securities		Taxes	
Mutual funds		Others	
Currency and coins			
Life insurance (cash value)			
Annuities (cash value)			
Income property			
Loans receivable			
Art objects (resale value)			
Business equity			
Vested pensions			
IRAs , KEOGH, 401(k)			
Home			
Automobiles, boats			
Clothing, jewelry			
Furniture			
Appliances			
Others			
Total Assets		Total Liabilities	
		Net Worth (Assets minus Liabilities)	

surprised after such an exercise because the calculated net worth turns out to be usually higher than expected.

One must be realistic when determining values for assets, especially furniture, art objects, and real estate holdings. Don't enter "may be" amounts; state the true current values, or in other words, the amount one could realize if the asset would be sold today. Case in point: if two similar houses in the neighborhood sold for 150,000 each but two repossesed homes have sold for 85,000 each, the true asset value of one's home would be the average of the four, namely 117,500, less real estate broker fees, taxes, and closing costs.

An appraiser should be consulted for rare art objects to determine their market value. The same applies to jewelry, coin, or other collections. Assets don't increase in value continuously; they adjust in accordance to the laws of supply and demand. It is better to use conservative values. Not what one thinks they should be, but values that realistically could be obtained if the asset would be put up for sale now.

Information on cash values of insurance policies and vested amounts in pension funds or annuities can be obtained from the trustees of these accounts. We have grouped the assets into two classes, namely, liquid and frozen assets. Liquid assets are all cash holdings and money market and savings accounts; in short, those assets that can be readily turned into cash on short notice. Frozen assets are those that cannot be quickly converted into cash; in short, items that might have to be sold at a substantial loss due to depressed market conditions or that could be sold only with a penalty attached to the transaction.

1.2 Determination of Monthly Cash Flow

Knowing the net worth leads to the next important step, namely, to find out how the money we earn is being spent on a monthly basis or what is known as calculating the monthly cash flow. Looking at the bank statement, many wonder where all the money has gone. Irrespective of the monthly income, there are thousands of households that live from pay check to pay check with nothing left over at the end of the month. Some cannot help it; their incomes are too low to have any hope of having something left over at the end of the month. But there are also others with sizable incomes spending their entire incomes every month, or worse, spending more than they earn. It is prudent to periodically sit down and analyze the monthly cash flow because it is part of the art of knowing how to manage money properly. To do this one sets up a calculating sheet as shown in Table 1.2.

Do not include interest and dividend payments from fund investments in the income column if they are automatically reinvested. Enter them only when these revenues are withdrawn to pay for living expenses. For rental, royalties, and other income, enter the amounts after expenses have been deducted.

Table 1.2 Determination of Monthly Cash Flow

Income		*Expenses*	
Salary (net)		A) Fixed	
Loans received		Rent	
Rental properties		Mortgage payments	
Social Security		Utilities	
Pension		Installment payments	
Gifts		Telephone, TV, cable	
Alimony		Leasing	
Dividends		Savings	
Interest		Health insurance	
Royalties		Taxes (Federal & state)	
Other income		Public transportation	
		Car (gas, maint., parking)	
		Child care	
		Memberships and fees	
		Books, education	
		Monthly services	
		Prescriptions	
		B) Variable	
		Food	
		Cleaning material	
		Personal hygiene	
		Entertainment	
		Postage, supplies	
		Pocket money	
		Newspapers	
		C) Reserves to be set aside	
		Clothes	
		Vacation	
		Car & home insurance	
		Home repairs	
		Gifts	
		Car (depreciation & repairs)	
		Dentist	
		Doctor visits	
Total Income:		Total Expenses:	
		Monthly Cash Flow (Income–Expenses) =	

Expenses fall into two groups: (a) fixed expenses that cannot be changed, and (b) variable expenses that can be adjusted. On purpose, we have placed savings close to the top of the fixed expenses column. Savings are the cornerstone in our method of building a sound financial future. Savings should be a fixed expense, not a variable one that one changes from month to month. It should be viewed as one of the most important expense items in any household.

What one has to strive for is a monthly positive cash flow, i.e., a situation where income exceeds expenses. Sustained negative cash flows indicate that one is living beyond one's means. In such situations it becomes necessary to reduce some of the variable expenses to bring the cash flow again to the positive side. Families going through periods of financial hardship might not be able to arrive at a positive cash flow for obvious reasons.

However, those with incomes sufficiently high to make the necessary adjustments, at this point, have to make some hard decisions.

Axiom I

Commit a fixed percentage of your monthly income for savings and make it a long term commitment.

Some might set this rate at 10 percent, others, with less income and more expenses, might be able to save only 5 percent. The percentage saved is not so important; of far more significance is that a commitment is being made to stick to a consistent savings plan for the long term. We know, it's not easy, especially not in this age of easy credit availability. We cannot promise something for nothing; becoming financially secure in the future requires hard commitments be made at a relatively young age. To make the monthly savings rate a fixed expense is one of the cornerstones for a secure financial future.

A frightful trend has developed in the past few years in matters of personal net income as applied to the middle class. This not only in the U.S. but in all other industrialized nations as well. Living costs have skyrocketed especially in the areas of health insurance, taxation, and public services, but the wages of the middle class have not kept pace with this escalation. Restructuring of many companies has led to massive lay-offs of workers and middle management personnel. The ones that have been lucky enough to keep their jobs often had to be satisfied with stagnant salary levels, and some even were forced to take wage reductions. With living costs increasing and incomes remaining the same or going down, the standard of living of many families decreased. In short, the purchase power of the middle class has been slowly and gradually eroded.

A nation's economy is predominantly governed by the spending habits of the middle classes. We exercise fiscal constraint in our spending, adapt to prevailing personal conditions, make spending cuts when needed, and manage our own personal affairs in such a manner that we can survive in the future. We have the right to ask the same from our politicians and from management at our workplace.

Think about it. How else can one become wealthy, short of a lottery or inheritance windfall, unless one has a clear-cut, disciplined approach to increase one's net worth. Setting aside money only when something is left over at the end of the month doesn't qualify as a legitimate savings plan. It allows for too much flexibility and opens the doors for splurges before savings are considered. By setting the savings as a fixed expense every month, it will become an obligation, a payment to be made just like any other due bill. Here, we are not talking about entering in a contractual agreement with an insurance company to purchase whole life insurance. Although these are being sold as "savings plans," I do not recommend them. They are relatively expensive and the investor doesn't have control over his savings when it comes to investments. The savings I am talking about in this book are funds the investor will control himself at all times.

Five or ten percent a month at first seems like a very large amount indeed. But it doesn't have to be; there are ways to cushion the impact. One might have a large debt load where installment payments eat all remaining funds at the end of the month. Obviously, in such situations one has to first get his financial affairs in order before one can think about setting aside money for savings and for investments.

It doesn't hurt to take the time to determine how much money is spent monthly on interest alone (not counting mortgage interest) on car, furniture, appliance, and credit card debt. One should then ask the question if it is possible to hold off on any future major purchases until the debt load is first completely eliminated. Are any of the items one plans to purchase in the near term absolutely necessary or could it wait until one can pay cash for these goods? Is it necessary to buy the flashy item with all the latest optional gismos when a less expensive model would do just as well?

The best way toward financial success is to pay cash for goods. If it becomes necessary to charge an item on your credit card, pay off the balance of the account within a month's time to avoid paying interest charges. When there is doubt the balance could be paid off right away, don't buy the item. One might even give serious consideration to paying off the mortgage on the house, especially in times when mortgage interest rates are on the high side. Houses don't increase in value for indefinite times; there is no guarantee one will be able to sell the home in the future at a profit. A job loss could result in a hard time keeping up with mortgage payments or, worse, being forced to sell the home at a time when the housing markets are depressed. Financial hardship due to overextended debt can befall everybody; nobody is immune. Living on credit doesn't lead to financial security.

Axiom II

With the exception of mortgage and car payments, pay cash for all purchases. When the credit card is used for purchases, pay off the balance within one month to avoid interest charges.

There will be many counterarguments to this axiom. In high inflationary times many financial advisors advocate using debt as leverage to combat the loss of purchasing power due to inflation. The technique of buying something now on credit and paying for it later with cheaper money has become a frightful trend worldwide. This mentality can have disastrous consequences when the economy turns from an inflationary to a deflationary environment. The savings and loan industry in the U.S. and Japan had to learn this truth the hard way. So did some portfolio managers who used margin and derivatives as a tool to increase the potential returns of the portfolio. Even governments are not immune from such pitfalls; their misguided fiscal and budgetary policies are often the leading causes for a financial crisis.

Irrespective of government actions, we still have the power to decide our own financial destiny. We can set our own goals and follow our own financial plan and should never place too much reliance on Social Security. There are no easy solutions, but surely there must be better ways than what we have now. A classical example of such a Social Security mess can be seen in Switzerland where retirees who have prudently saved all their lives for a comfortable retirement are being excessively taxed.

1.3 Amount Needed for Retirement

How much of a nest egg is required to attain financial security without having to depend on Social Security payments once we retire from the work force? What amount is required to retire comfortably? Is there a limit to the size of a portfolio at which an investor could stop saving and start spending the money he has accumulated? There is, but, the amount needed is much higher than what most investors expect.

The following formula (1.3.1) will allow the reader to calculate the amount required to fulfill the dream of a comfortable retirement and is based on the assumption of the investment portfolio yielding a simple annual return of 5%. Please note that we show the formula style used for computer programming for any spreadsheet program in this and all subsequent formulas in this book.

$$P = 1/\left(\left(0.05^{*}\left((1.05)^{\wedge}n\right)\right)/\left(a^{*}\left(\left((1.05)^{\wedge}n\right)-1\right)\right)\right) \qquad 1.3.1$$

where P = principal amount needed
n = number of retirement years
a = annual withdrawal in retirement

One should always expect to live at least to age 90 in such calculations. Table 1.3 shows that someone retiring at age 65 and budgeting withdrawals of 50,000 per year, needs a portfolio size of 704,697 at the start of his retirement (retirement years = 90 – 65 = 25). In another example, someone planning early retirement at age 55 (retirement years: 90 – 55 = 35) with annual withdrawals of 60,000 would have to possess a nest egg of 982,452 to make his dreams possible. Even if this wealth is attained, the question remains if it is sensible to retire at such an early age when an individual is still healthy and relatively happy with his current job. Playing golf or going fishing every day or travelling to every corner of the world might become a bore after a while. Watching television or surfing the Internet all day long for 35 years of retirement is also not the answer.

An excellent method to start saving for the future is through an employer's savings plan (401 k) when available. Such plans carry a triple advantage: (a) they force an individual to a rigid percent savings on a monthly basis, (b) provide an up-front guaranteed return because the employer usually

Table 1.3 Amount Needed for Retirement Based on 5% Annual Return

Annual Withdrawal	*Total Years of Expected Retirement*						
	5	*10*	*15*	*20*	*25*	*30*	*35*
15000	64942	115826	155695	186933	211409	230587	245613
20000	86590	154435	207593	249244	281879	307449	327484
25000	108237	193043	259491	311555	352349	384311	409355
30000	129884	231652	311390	373866	422818	461174	491226
35000	151532	270261	363288	436177	493288	538036	573097
40000	173179	308869	415186	498488	563758	614898	654968
45000	194826	347478	467085	560799	634228	691760	736839
50000	216474	386087	518983	623111	704697	768623	818710
55000	238121	424695	570881	685422	775167	845485	900581
60000	259769	463304	622779	747733	845637	922347	982452
65000	281416	501913	674678	810044	916106	999209	1064323
70000	303063	540521	726576	872355	986576	1076072	1146194
75000	324711	579130	778474	934666	1057046	1152934	1228065
80000	346358	617739	830373	996977	1127516	1229796	1309936
85000	368006	656347	882271	1059288	1197985	1306658	1391807
90000	389653	694956	934169	1121599	1268455	1383521	1473677
95000	411300	733565	986068	1183910	1338925	1460383	1555548
100000	432948	772173	1037966	1246221	1409394	1537245	1637419

matches a certain percentage of the contribution, and (c) since such plans are in most countries tax-deferred, one is forced to leave these savings to accumulate until age 59.5. In short, it introduces financial discipline into everyday life.

The examples given above are indeed large numbers and seem to be out of reach for most people. This is the reason why it is so important to plan for retirement when one is still relatively young. The most important factor that makes a portfolio grow is time. The power of compounding is most often the key factor behind a financial success story.

Still having doubts or apprehensions about making a savings commitment? Then compare the following two cases:

Case A: The Saver:

Saves $250 per month. Has no debt or interest to pay and lets the savings and interest accumulate at an annual rate of 7%.

Net worth after 25 years: $202,518

Case B: The Spender Using Credit:

Has no savings and pays $250 per month interest for loans and credit card charges.

Total loss after 25 years: $60,000

Total difference between the two extreme cases is $262,518 over a period of 25 years.

There is no need to become paranoid in the beliefs of an imminent disaster or major collapse of the economy. No need to convert all our assets into gold, move to rural areas, and start to grow our own food. Such dire predictions make good copy and sell a lot of books, but are usually detrimental to one's financial well-being. Nevertheless, should a crisis develop in the future, there will be time to make appropriate adjustments in the portfolio to prevent a substantial loss of capital. We will cross that bridge when we come to it, not before.

Axiom III

Have 5 month's worth of living expenses at all times as reserves in a money market fund.

The five-month living expense reserves are viewed as the security blanket for potential major losses of income that could occur in any household. This money is preferrably deposited in a money market account and is part of our overall investment portfolio.

Money deposited in money market accounts can be withdrawn in an emergency at short notice. Large unexpected doctor or dentist bills will also be paid out of this emergency fund. One should never view the credit line on charge accounts as an emergency blanket. Charging such large amounts with the credit card and paying off the debt in monthly installments at 18% interest is a sure way to lose sight of the investment goals. This emergency fund should not be a source to pay for vacations or for major appliance purchases.

1.4 What Kind of Life Insurance?

When an investor carries the financial responsibility for the entire family, he must consider some form of life insurance to provide security for the dependents. Unfailingly, he will sooner or later be approached by a salesperson trying to sell him whole life or universal life insurance. Likely, the financial advisor will be able to produce all kinds of fancy charts showing how attractive such a life insurance policy is. He will talk about the cash value such insurance will build over time and at the same time provide financial security for the dependents should an accident happen. Most of the time the salesman will try to have the client commit himself for an insurance that runs into the tens of years because long term contracts generate more commissions for the salesman. Don't go for it; it's not as good an investment as he tries to describe it. More often than not, it is a losing proposition when one takes inflation into account.

When one has children or is the sole provider of income in the family, then by all means security should be provided by a life insurance but this should be term life insurance. This type of insurance doesn't tie insurance together with the investment plan. Term life insurance is a straight insurance expense and doesn't accumulate cash value. It is much cheaper than whole life insurance. Thus, when a salesman keeps on pressuring for whole life and shows little interest in discussing term life — terminate the discussion and show him the door.

Expert opinion should be solicited before any insurance policy is signed because insurance salesmen have the tendency to wrap all kinds of extra benefits into a policy, extra benefits that are expensive and not needed. They are made to look important by the salesperson because they will generate more commissions into his pockets.

1.5 Systematic Investment Plans

Some mutual fund companies offer investors systematic savings plans where an investor can enter into a contract to make monthly deposits of $50, $100, or more for a given length of time such as 15 or 25 years. These plans force an investor to adhere to this savings regimen for a long period of time.

Premature termination of such a contract can be costly because most of the sales fees are deducted in the first year of the plan's life. The following example shows clearly how these sales charges impact the returns of such plans in the first 5 years. Not only that, annual returns are still governed by prevailing market gyrations and the fund manager's skills to produce reasonable returns. Naturally, no great results can be obtained if 50% of the first year's deposits are taken away in form of sales charges and fees. This example doesn't reflect returns that can be expected from each one of such funds. It only serves to demonstrate what an investor is getting into when he commits to such a plan. I don't recommend such a plan because we want to have full control of our own investment strategy and don't want to rely for 15 years on one fund manager to produce the returns we need.

Example of Fees and Returns in an Investment Plan

Year	Cumulative Invested	Annual Fees	Cumulative Distributions	Cumulative Value	Cumulative % Gain
1	600	313	33	268	–55
2	1200	34	94	906	–25
3	1800	34	475	2223	24
4	2400	34	868	2854	19
5	3000	34	1309	3758	25
10	6000	34	5518	10359	73
15	9000	34	24989	36432	305

Conclusion: if any such contractual savings plan is contemplated to be purchased, make sure to stay with the plan for the full duration. Premature canceling of this contract, especially if it happens in the first 3 years, is a losing proposition.

A few more things are required to bring our financial house in order:

- Estate planning: A will has to be drawn up. Also consider the possibility of setting up a living trust. Don't wait until old age; do it irrespective of age. Have a legal expert view this document to make sure it is properly worded and legal. A properly drawn will can save the heirs large sums of money and can speed the process of distributing your assets primarily to the persons you want it to go — not the government or some lawyers. A living trust can also save probate and lengthy legal fights among survivors. The important thing here is to get these documents drawn up by experts familiar with the laws in your area of residence. The fees for such a service are well spent and will give you and your relatives peace of mind.

- The latest net worth table with names, addresses of investment accounts, safe deposit box locations, and of persons to contact in case of death should also be attached to this will.
- An inventory of all household goods, including their estimated resale value should be made, or better, be recorded on video. These records should be kept in a fireproof safe.
- Confide in your spouse, children, or a very close person to let that person know where these important documents are stored.
- When considerable assets and/or sizable investment portfolios are at stake, consideration should be given toward lessening the tax burden of the heirs. It then makes sense to consult a tax expert for estate planning.

There is nothing morbid about these steps and there are many ways trust accounts can be set up to keep the wealth in the family. It is only logical to consider all these possibilities ahead of time instead of letting a sizable portion of your estate go to the government in the form of inheritance taxes later.

Now, let's assume the financial house is in order. Our financial situation has been analyzed. Debt has been reduced to a minimum, a five-month living expense emergency account has been started, and a savings plan has been decided upon to reach a given goal after x-years. The road to personal financial independence has thus been paved. Good. — But, now come the headaches.

The question now becomes, how can one make one's portfolio generate a true return after taxes and inflation year after year? How, in what, when, and where should the portfolio assets be invested? Answers to all these questions will be given in the remaining chapters of this book.

2 Growth of a Portfolio

2.1 Mutual Fund Returns

Open any financial paper today and you'll likely find an advertisement where an investment advisor or mutual fund company claims staggering returns over the past five or ten years. For example, our attention might be drawn to the headline: "Dinero Equity Fund: 216% total return for the past 10 years."

At first, this looks like a terrific return. After all, that amounts to an annual 21.6%, the uninitiated might conclude. This is clearly a wrong conclusion as we will explain shortly. The wording in these advertisements mostly attempts to make the reader believe that these high returns are the direct results of the advisor's or fund manager's superior investment skills. Nothing could be further from the truth. This is one of many games the investment community plays to attract the attention of the small, individual investor.

"Total Return" means a given amount of money was invested in a fund some years ago and all interest, dividends, and capital gains were reinvested, i.e., never during the entire period did the investor take money out of the fund. In other words, the investor took advantage of the time factor to let his investment grow by the power of compounding. Total return is also often referred to as compounded return.

2.2 Compounded Interest

Table 2.1 shows the factors to be used to calculate the growth of an investment over a given time when all the interest earned is reinvested (compounded). This table can thus be used to quickly determine the future value of an investment. For example, an initial portfolio of 75,000, with an assumed annual return of 8% will have a value of 349,575 after 20 years (75,000 * 4.661 = 349,575).

Table 2.1 Compound Interest Factors

	Annual Simple Interest (percent)											
Years	*3.0*	*3.5*	*4.0*	*4.5*	*5.0*	*6.0*	*7.0*	*8.0*	*9.0*	*10.0*	*11.0*	*12.0*
1	1.030	1.035	1.040	1.045	1.050	1.060	1.070	1.080	1.090	1.100	1.110	1.120
2	1.061	1.071	1.082	1.092	1.103	1.124	1.145	1.166	1.188	1.210	1.232	1.254
3	1.093	1.109	1.125	1.141	1.158	1.191	1.225	1.260	1.295	1.331	1.368	1.405
4	1.126	1.148	1.170	1.193	1.216	1.262	1.311	1.360	1.412	1.464	1.518	1.574
5	1.159	1.188	1.217	1.246	1.276	1.338	1.403	1.469	1.539	1.611	1.685	1.762
6	1.194	1.229	1.265	1.302	1.340	1.419	1.501	1.587	1.677	1.772	1.870	1.974
7	1.230	1.272	1.316	1.361	1.407	1.504	1.606	1.714	1.828	1.949	2.076	2.211
8	1.267	1.317	1.369	1.422	1.477	1.594	1.718	1.851	1.993	2.144	2.305	2.476
9	1.305	1.363	1.423	1.486	1.551	1.689	1.838	1.999	2.172	2.358	2.558	2.773
10	1.344	1.411	1.480	1.553	1.629	1.791	1.967	2.159	2.367	2.594	2.839	3.106
11	1.384	1.460	1.539	1.623	1.710	1.898	2.105	2.332	2.580	2.853	3.152	3.479
12	1.426	1.511	1.601	1.696	1.796	2.012	2.252	2.518	2.813	3.138	3.498	3.896
13	1.469	1.564	1.665	1.772	1.886	2.133	2.410	2.720	3.066	3.452	3.883	4.363
14	1.513	1.619	1.732	1.852	1.980	2.261	2.579	2.937	3.342	3.797	4.310	4.887
15	1.558	1.675	1.801	1.935	2.079	2.397	2.759	3.172	3.642	4.177	4.785	5.474
16	1.605	1.734	1.873	2.022	2.183	2.540	2.952	3.426	3.970	4.595	5.311	6.130
17	1.653	1.795	1.948	2.113	2.292	2.693	3.159	3.700	4.328	5.054	5.895	6.866
18	1.702	1.857	2.026	2.208	2.407	2.854	3.380	3.996	4.717	5.560	6.544	7.690
19	1.754	1.923	2.107	2.308	2.527	3.026	3.617	4.316	5.142	6.116	7.263	8.613
20	1.806	1.990	2.191	2.412	2.653	3.207	3.870	4.661	5.604	6.727	8.062	9.646
21	1.860	2.059	2.279	2.520	2.786	3.400	4.141	5.034	6.109	7.400	8.949	10.80
22	1.916	2.132	2.370	2.634	2.925	3.604	4.430	5.437	6.659	8.140	9.934	12.10
23	1.974	2.206	2.465	2.752	3.072	3.820	4.741	5.871	7.258	8.954	11.03	13.55
24	2.033	2.283	2.563	2.876	3.225	4.049	5.072	6.341	7.911	9.850	12.24	15.18
25	2.094	2.363	2.666	3.005	3.386	4.292	5.427	6.848	8.623	10.83	13.59	17.00

Although Table 2.1 shows the appropriate factors to use when the interest is compounded annually, other similar tables can be made for investments that compound daily, monthly, or quarterly. The interest (i) must then be calculated for the given period (n) according to the following formula:

$$\text{Compound Interest Factor} = (1+i)^n \qquad 2.1$$

where i = Interest rate in given period
n = Number of periods

Now let us go back and see how the manager of the Dinero Equity Fund made out in the earlier given example. 216% total return in 10 years corresponds to a compound interest factor of 216/10= 2.160. A glance at Table 2.1 tells us that the manager has achieved a simple annual return of just under 8%. In truth then, the manager produced an 80% return in ten years and the investor can take credit for the other 136% because he reinvested all the fund distributions in form of dividends and capital gains, i.e., never in the ten years did the investor take any money out of the fund. In stating their performance in terms of total return, fund managers decorate themselves with garlands that rightfully should not hang around their necks. But again, 216% looks a lot better than 80% and certainly attracts more attention.

It gets worse. For the same period, the market index by which performances of equity fund managers are measured has achieved a much higher average annual simple return during this ten year period. In short, the fund manager has grossly underperformed the market average. In addition, the fund charged a front end load and debited the fund 1.5% of assets per year for management fees. With total assets of 500 million, these fees amounted to 7.5 million — not over the 10-year period — but every year. Some management, some fees. This is not an isolated case; underperformance when compared to a market index, seems to be more the norm than the exception among fund managers. Later in this book we will show a method for filtering out the good from the mediocre fund managers.

In fairness, it has to be pointed out that there are many skilled portfolio managers that consistently have shown superior performance results. For example, Peter Lynch, the former manager of the Fidelity Magellan Fund, achieved a total return of 1168% in a ten-year period that included the 1987 market crash. Truly a remarkable performance considering that this is equivalent to an average annual simple return of 29%.

Sometimes fund performance is also shown by the growth of an initial investment of 10,000 over a given period of time. For example, an investment of 10,000 grown to 25,900 in ten years, according to Table 2.1 represents a compound interest factor of 25,900/10,000 = 2.59 and a simple annual average return of 10%.

The investment technique described herein combines the benefits of compounding with regular monthly savings to reach the goal of financial security. An investment very seldom will show significant growth in value over time unless one takes advantage of the time compounding factor and commits oneself to regular and consistent savings in periodic and fixed intervals. Table 2.2 shows the growth of a portfolio over time when monthly savings of 250 are deposited into the account and one lets the distributions compound at a given interest rate. For example, 250 consistently invested every month, at 9% interest, will have a value of 282,283 after 25 years. Table 2.3 applies for monthly savings of 350, both tables having been calculated based on the formula

$$V = \frac{S}{i}\left(\left(1+i\right)^{n+1} - \left(1+i\right)\right) \tag{2.2}$$

for computer programming: $V = (S/i) * ((1+i) \wedge (n+1) - (1+i))$

where V = final value after n periods
S = amount of savings per period
i = interest earned per period
n = number of periods

2.3 Fund Management Fees

The topic of management fees now has to be addressed. When the total fees of a fund are set at 2%, this means the return of the fund will be reduced every year by this percentage, irrespective if the fund had a dismal or superior performance. These fees come directly out of the fund assets and are therefore reflected in a lower Net Asset Value (NAV). Without these fees, the performance of the fund would have been 2% higher for any given year. Now, let's go back and take a quick look at the Compound Interest Factors (Table 2.1) and assume the following investment condition:

Table 2.2 Portfolio Growth with 250 per Month Deposits

	Annual Interest→	3.0	4.0	5.0	6.0	7.0	8.0	9.0	10.0	11.0	12.0
	Monthly Interest→	0.0025	0.0033	0.0042	0.0050	0.0058	0.0067	0.0075	0.0083	0.0092	0.0100
Years	Number of Deposits										
1	12	3049	3066	3083	3099	3116	3133	3150	3168	3185	3202
2	24	6191	6257	6323	6390	6458	6527	6596	6667	6738	6811
3	36	9429	9577	9729	9883	10041	10201	10365	10533	10703	10877
4	48	12765	13033	13309	13592	13883	14181	14488	14803	15126	15459
5	60	16202	16630	17072	17530	18003	18492	18997	19521	20062	20622
6	72	19744	20373	21028	21710	22420	23160	23930	24732	25568	26439
7	84	23394	24269	25187	26149	27157	28215	29325	30490	31712	32995
8	96	27155	28324	29558	30861	32237	33690	35226	36850	38566	40382
9	108	31030	32543	34153	35863	37683	39620	41681	43876	46214	48705
10	120	35023	36935	38982	41175	43524	46041	48741	51638	54747	58085
11	132	39137	41506	44059	46814	49786	52996	56464	60213	64267	68654
12	144	43377	46263	49396	52800	56501	60528	64911	69685	74889	80563
13	156	47745	51213	55006	59156	63702	68685	74151	80150	86740	93983
14	168	52247	56365	60902	65904	71423	77519	84257	91710	99962	109104
15	180	56885	61728	67101	73068	79703	87086	95311	104481	114714	126144
16	192	61664	67308	73616	80674	88581	97448	107402	118589	131174	145345
17	204	66589	73116	80465	88749	98100	108669	120628	134175	149538	166980
18	216	71664	79161	87664	97322	108308	120822	135094	151392	170027	191360
19	228	76893	85452	95232	106424	119254	133983	150917	170412	192888	218831
20	240	82281	91999	103187	116088	130991	148237	168224	191424	218393	249787

Table 2.2 (continued) Portfolio Growth with 250 per Month Deposits

	Annual Interest→	3.0	4.0	5.0	6.0	7.0	8.0	9.0	10.0	11.0	12.0
	Monthly Interest→	0.0025	0.0033	0.0042	0.0050	0.0058	0.0067	0.0075	0.0083	0.0092	0.0100
Years	Number of Deposits										
21	252	87833	98813	111548	126347	143577	163674	187155	214636	246850	284669
22	264	93553	105905	120338	137239	157072	180392	207862	240279	278601	323974
23	276	99448	113285	129577	148803	171543	198497	230511	268607	314025	368264
24	288	105522	120967	139289	161080	187060	218106	255285	299901	353548	418172
25	300	111781	128961	149498	174115	203699	239342	282383	334473	397645	474409
26	312	118230	137281	160229	187953	221541	262340	312022	372664	446845	537778
27	324	124875	145940	171509	202645	240672	287247	344443	414854	501739	609184
28	336	131722	154951	183366	218243	261187	314222	379904	461462	562984	689646
29	348	138778	164330	195830	234803	283184	343436	418692	512951	631317	780313
30	360	146048	174091	208932	252384	306772	375074	461119	569831	707557	882478
31	372	153540	184249	222703	271050	332065	409338	507525	632668	792620	997601
32	384	161259	194822	237180	290867	359186	446446	558285	702084	887525	1127324
33	396	169213	205825	252397	311907	388268	486634	613806	778769	993414	1273500
34	408	177409	217276	268393	334244	419452	530158	674536	863484	1111555	1438213
35	420	185854	229194	285207	357958	452890	577294	740962	957069	1243368	1623817
36	432	194557	241598	302881	383136	488746	628342	813620	1060454	1390434	1832960
37	444	203523	254507	321459	409866	527194	683627	893093	1174665	1554518	2068628
38	456	212763	267941	340988	438245	568421	743501	980022	1300836	1737590	2334184
39	468	222283	281924	361516	468374	612628	808345	1075105	1440218	1941847	2633419
40	480	232094	296475	383095	500362	660031	878570	1179108	1594195	2169740	2970605

Table 2.3 Portfolio Growth with 350 per Month Deposits

	Annual Interest→	*3.0*	*4.0*	*5.0*	*6.0*	*7.0*	*8.0*	*9.0*	*10.0*	*11.0*	*12.0*
	Monthly Interest→	*0.0025*	*0.0033*	*0.0042*	*0.0050*	*0.0058*	*0.0067*	*0.0075*	*0.0083*	*0.0092*	*0.0100*
Years	*Number of Deposits*										
1	12	4269	4292	4316	4339	4363	4387	4410	4435	4459	4483
2	24	8668	8759	8852	8946	9041	9137	9235	9334	9434	9535
3	36	13200	13408	13620	13836	14057	14282	14511	14746	14984	15228
4	48	17870	18246	18633	19029	19436	19854	20283	20724	21177	21642
5	60	22683	23282	23901	24542	25204	25888	26596	27329	28086	28870
6	72	27642	28523	29440	30394	31388	32424	33502	34625	35795	37015
7	84	32751	33977	35261	36608	38020	39501	41055	42685	44397	46193
8	96	38016	39653	41381	43205	45131	47166	49317	51590	53993	56534
9	108	43442	45561	47814	50209	52757	55468	58354	61426	64700	68188
10	120	49032	51709	54575	57645	60933	64458	68238	72293	76646	81319
11	132	54792	58108	61683	65539	69701	74194	79050	84298	89974	96115
12	144	60727	64768	69154	73920	79102	84739	90876	97560	104844	112788
13	156	66843	71698	77008	82819	89183	96159	103811	112210	121436	131576
14	168	73145	78912	85263	92266	99993	108527	117959	128394	139947	152746
15	180	79639	86419	93941	102295	111584	121921	133435	146273	160600	176602
16	192	86330	94232	103063	112944	124013	136427	150363	166025	183643	203482
17	204	93225	102363	112651	124249	137341	152137	168879	187844	209353	233772
18	216	100329	110826	122730	136251	151632	169150	189131	211949	238038	267904
19	228	107650	119633	133325	148994	166956	187576	211283	238577	270043	306364
20	240	115193	128799	144461	162523	183388	207532	235514	267994	305751	349702

Table 2.3 (continued) Portfolio Growth with 350 per Month Deposits

	Annual Interest→	*3.0*	*4.0*	*5.0*	*6.0*	*7.0*	*8.0*	*9.0*	*10.0*	*11.0*	*12.0*
	Monthly Interest→	*0.0025*	*0.0033*	*0.0042*	*0.0050*	*0.0058*	*0.0067*	*0.0075*	*0.0083*	*0.0092*	*0.0100*
Years	*Number of Deposits*										
21	252	122966	138339	156168	176886	201008	229143	262017	300491	345591	398536
22	264	130975	148267	168473	192135	219901	252548	291006	336391	390041	453564
23	276	139227	158600	181408	208324	240161	277896	322715	376050	439635	515570
24	288	147731	169353	195005	225512	261885	305348	357399	419862	494967	585441
25	300	156493	180545	209297	243761	285179	335078	395336	468262	556703	664172
26	312	165522	192193	224320	263134	310157	367276	436831	521729	625583	752889
27	324	174825	204315	240113	283703	336941	402146	482220	580796	702434	852858
28	336	184411	216932	256713	305540	365662	439911	531866	646047	788178	965505
29	348	194289	230062	274162	328724	396458	480810	586169	718132	883844	1092438
30	360	204468	243727	292504	353338	429481	525103	645566	797764	990580	1235470
31	372	214956	257949	311785	379470	464891	573073	710535	885735	1109667	1396642
32	384	225763	272750	332052	407214	502860	625024	781599	982917	1242536	1578254
33	396	236898	288155	353356	436669	543575	681288	859328	1090276	1390779	1782899
34	408	248373	304187	375750	467941	587232	742221	944350	1208877	1556177	2013499
35	420	260196	320872	399289	501142	634046	808211	1037347	1339897	1740715	2273344
36	432	272379	338237	424033	536390	684244	879679	1139068	1484636	1946608	2566144
37	444	284933	356309	450043	573813	738071	957078	1250330	1644532	2176326	2896079
38	456	297868	375118	477383	613543	795789	1040902	1372030	1821170	2432627	3267858
39	468	311197	394693	506123	655724	857679	1131683	1505147	2016305	2718586	3686787
40	480	324931	415065	536333	700507	924044	1229998	1650751	2231873	3037637	4158847

3 Types of Fund Investments to Consider

A list of groups of investment possibilities available to investors could likely fill three or four pages. This is one of the advantages of living in countries where the capital markets have fully matured and can offer an investor just about everything where an opportunity exists to make some money (and we might add, to lose one's life savings). Yes, the system also has its disadvantages. A highly competitive market also has considerable excesses in advertising hype and can thus easily confuse the small investor. Worse, it could lure him into investments he should never have thought about entering in the first place.

3.1 Portfolio Diversification

At some time or another, every investor will have to address himself to the subject of diversification. The phrase, "Don't put all your eggs into one basket," has been hyped and beaten around so much with the result that many investors have over-diversified their portfolios and thus have to be satisfied with mediocre results.

Over-diversification is quite common with investors who have accumulated a sizable portfolio. One should never forget that adding a new account or fund to the portfolio will invariably involve some management and transfer fees that depress the investment returns. Although such fees are necessary considering the paperwork financial institutions are required by law to complete, these charges can accumulate to a sizable amount over a period of say twenty or thirty years. We repeat again, it shall not be our purpose to make the investment houses rich; our goals are to keep profits in our own portfolio.

As already indicated, our investment strategy does not have any room for leveraged, highly speculative investments. Options and futures funds and buying investments on margin are out. Investing in real estate and income properties when it involves borrowing to finance the initial purchase, is also not open for discussion because it is speculative in nature and poses risks we do not want to take with our hard earned money. We don't want to be daredevils with objectives of becoming filthy rich. All we want to do is to keep what we have, to maintain the purchasing power of our portfolio, and to get an additional reasonable premium return for our efforts in investing in stock and bond funds.

Most individual investors have equity accumulated in their own homes and look at this equity as part of their investment portfolio. The owner's residence should not be viewed as such. Neither should one classify his car, jewelry, or furniture as investments. These are part of the Net Worth and should not be grouped into the class of investments. The investment portfolio that will take care of our financial needs after retirement will be predominantly made up of our savings in mutual funds. What we will do with the equity in our own residence will become an investment issue only at the time when the home is sold and the proceeds are deposited into our savings pool.

3.2 Type of Funds for our Investment Strategy

Investments considered in our strategy are

Money market funds
Aggressive growth funds
Growth-income funds
Intermediate and long term bond funds
High income bond funds
Foreign equity and bond funds.

3.3 The Basis of our Investment Strategy

How many and which types of funds we will include in our portfolio will depend on

1. The size of our investment portfolio
2. The number of years remaining until retirement

3. The personal risk level of the investor compared to the risk status of the investment portfolio as a whole.

These are the three cornerstones upon which we structure our investment portfolio. The personal risk level depends on items (1) and (2) and the portfolio risk level is determined from our method of asset allocation analysis. These items will be discussed in detail later since they are considered the most important factors that govern the success of our mutual fund investment strategy. Ignoring or failing to give consideration toward these factors will lead to an incomplete investment concept. Worse, it could result in a portfolio mix that is too conservative or too speculative for a given investor's personal financial situation. In short, our strategy allows an investor to customize his investment portfolio to his exact personal needs and goals.

There will be times when an investor will have all of the above listed types of funds in his portfolio, but there will also be times when our strategy demands most of the assets to be invested in only one or two types of funds. Depending on the size of the portfolio, assets should preferrably be diversified in funds as shown in Table 3.1.

Table 3.1 Fund Portfolio Diversification

	Portfolio Size (000'S)			
	>25	*>75*	*>150*	*>300*
Money market fund	1	1	1	1
Growth-income funds	1	2	2	3
Growth funds	1	1	2	3
Bond funds (interm-long)	1	1	1	2
High income bond funds				1
Foreign funds			1	2
Total number of funds	4	5	7	12

This table doesn't show how much capital should be invested in each of the funds. This question will be answered later when our asset allocation technique is discussed.

I recommend primarily mutual funds for investments because funds themselves are composed of a multitude of individual securities and therefore already diversified. Once the portfolio has reached a value in excess of $300,000, an investor might consider investing in individual securities instead of funds. This would however require the portfolio to contain many more components as shown in Table 3.2.

Table 3.2 Individual Stock Portfolio Diversification

	Portfolio Size (000's)		
	>150	*>350*	*>500*
Money market account	1	1	1
Security trading account	1	1	1
Growth-income stock	4	5	6
Growth stock	3	5	7
Government bonds	1	2	3
Corporate bonds	3	4	5
Foreign bonds	1	2	3
Foreign stocks	4	5	6
Total individual holdings	18	25	32

An investor, selecting individual securities on his own, would have to keep track of at least 18 individual stocks and bonds. This requires considerable time most investors don't have. Besides, this would also require specialized knowledge of bond and stock selecting techniques, attributes most investors with small portfolios have not yet acquired. This clearly speaks in favor of investing in mutual funds where all the analytical work is done by the fund managers and analysts.

The main purpose of diversification is to reduce the overall risk of a portfolio. Defining investment risk is a highly questionable subject. The author has concluded that the standard mathematical method used today to measure risk, namely, the Beta, Alpha, and Correlation (R-squared) coefficients of an investment, when applied to mutual fund investments, is fundamentally flawed despite its wide acceptance in the academic and financial community. Just because about every professor subscribes to this theory doesn't necessarily mean the theory is sound. The concept of having an equity mutual fund with a Beta of 0.70 lose "only" 35% when the market goes down 50%, doesn't make any sense to any serious investor. To assign a less risky tag to such a fund doesn't seem to be the correct term. In Chapter 4 a method is shown to express mutual fund investment risk in more realistic terms.

The fact is some types of investment funds react differently in a down market than in an up market when compared to the market average. The converse is also true. In short, each investment has its own specific down and up side risk. The two risks are not mutually equal as the Beta theorists want us to believe.

The next axiom specifies the funds we are looking for.

Axiom V

1. Invest only in open-end funds,
2. Buy no-load funds or buy funds with 3% or less load,
3. No hidden 12b-1 charges,
4. Buy funds with management fees below 1% for domestic and below 2% for foreign funds.

Closed-end funds are not considered in our strategy because they act and trade more like individual stocks.

To facilitate adjustments in portfolio structures and to make accounting procedures easier, it is recommended to purchase all the funds from the same mutual fund company. To qualify, the company must

1. offer a family of funds and must have at least five different types of funds available for each of the various types in Table 3.1 listed investment alternatives
2. offer 24-hour telephone exchange or internet access privileges.

Mutual funds, for our investment purpose, should preferably not be purchased from banks, brokers, insurance, or investment advisors. Funds thus purchased often carry a heavy load of up to 8%. It doesn't make financial sense to pay someone a high entry fee that mostly represents the commission for the sales agent and broker and by no means pledges superior performance. Most of these funds perform as good or as bad as no-load funds. Take for example an investor who buys $50,000 worth of shares in a fund that has an 8% load up front. Immediately $4000 will be taken out for the load with no apparent benefits other than for the sales agent or brokerage house — a staggering sum of money for a few minutes talking to a broker to execute the trade. Don't forget, there still will be annual management fees charged that are in addition to the load. As common sense investors, we do not want to stuff somebody else's pockets when the performance is likely not better than a fund charging no entry fee. Then therc are also funds that let you in for free but won't let you go unless you pay an exit fee. Like entering a long tunnel free, knowing there will be a toll booth at the other end.

Beware of fancy performance charts these sales representives can present; they can often dazzle the uninitiated. It doesn't need a super talent of a fund manager to show good looking charts in a raging bull market. Ask these so-called investment advisors to show you the charts when the market was so-so or down. Here is where the good managers will shine.

We have specified six types of fund investments suitable for our strategy. Since we invest exclusively in mutual funds, we shall not discuss the merits of investing in treasury bills or CDs although they are legitimate alternatives.

Again, the list above doesn't imply we will at all times be invested in each of these types of funds. But, we will evaluate each type for its merits in order to be included in our portfolio as a potential contributor to increased returns and reduced overall portfolio risk. At any given time, there will be some types or groups exerting a positive and some a stagnant impact on portfolio performance. By mixing groups of funds we will be able to implement our investment strategy that entails primarily the concept of preservation of capital with a reasonable return. It doesn't mean each of the selected individual funds at all times will deliver the desired return because we never know what the future will hold, but each fund has to make a positive contribution toward the portfolio return as a whole over a period of say three years.

With a sound investment strategy, there will be no good and bad investment times. Our active investment strategy calls for identifying the oportunities and switching in and out of funds in infrequent intervals, perhaps adjusting the portfolio once or twice a year. A passive investment style is also not in our books because we cannot "sit" on an investment that lost considerable value because, as the erroneous saying goes, "It's only a paper loss; it will come back again." Likewise, staying with an investment that performed well in the past under the assumption that it will continue to be a good performer in the future doesn't fit our strategy either. A successful investor will know when to buy in bad and when to sell in good times. Everybody wants to buy low and sell high, but few can gather the courage to go contrary to the crowd. Emotions and the urge to follow the crowd often overrule an investor's sound judgement. One should not feel bad about it; money managers, investment advisors, and fund managers fall into this trap a lot of times, too.

The Fuzzy Logic strategy described herein lets the air out of wishful thinking, gazing into the future, and all the hype that prevails in today's market. As we will describe later in detail, we focus on the portfolio risks as a whole, compare it to our own risk level, and make adjustments when the two are too far apart from each other. It sets us on solid footing to deal with reality and allows us to do what we have in mind first and foremost, namely, make money the sensible way.

3.4 Money Market Funds

There is a widespread misconception about money market funds. All too often, they are advertised as being a safe, risk-free investment. The main selling point being made is that money market funds maintain a constant NAV (Net Asset Value) at all times and thus, the saying goes, the investment capital is not at risk. The sales argument that money market funds have never lost principal, i.e., have never traded below unity, is wishful thinking along the line of thought that if it didn't happen in the past, it cannot happen in the future. That's like saying: having never had a case of cancer for generations in the family will guarantee that we cannot be stricken by this disease in the future. The fact is that some money market funds invest part of the total assets in riskier short term instruments of shaky corporations, banks, or insurance companies to enhance the yield of the fund. If a cataclysmic financial disaster should befall the financial world and mass bankruptcies become the norm, many money market funds will then not be able to maintain the constant NAV of 1.00. Its value could go to a discount and thus erode the principal. Chances of this happening are remote, but we do not know what the future holds.

Some recommend using money market funds as a temporary short term "parking place," i.e., only until such time when more favorable situations return to the stock or bond markets. This line of thinking appears to run parallel with many a financial advisor's contention that high returns can only be achieved by being predominantly invested in equities. The fact is that a money market fund can be an investor's best friend when things go south in the stock and/or bond market. More important, in our strategy we use money market funds to control the overall risk of the portfolio as a whole. We do not view money market funds as a temporary "safety valve" to be abandoned as soon the stock and bond markets are booming again. No, we place as much importance and attention on these funds as any other type of fund investment we hold. Money market funds are a permanent and integral part of our investment strategy.

Money market funds invest in interest bearing short term securities. Most of the securities have a maturity of less than 120 days. Interest earned in money market funds is generally credited and compounded daily to the investor's fund account. Interest paid out varies according to prevailing interest rate trends. It is important to note that a typical money market fund is not federally insured. There are some insured money market funds available for purchase, but their yield is lower and not worth considering for our portfolio strategy unless we are in a high tax bracket.

The choice narrows down to two types of money market funds, namely, taxable and tax-free funds. For investors in a high tax bracket and those with sizable sums in money market accounts, it might be advantageous to use tax-free funds. Tax-free funds should be considered only if their yield is higher than the calculated after-tax yield of taxable funds.

$$Tf = Tx * (1 - b) \qquad 3.1$$

where Tf = minimum yield in tax-free fund required
Tx = current yield of taxable money market funds
b = investor's tax bracket (expressed as a decimal)

Thus, an investor in the 33% tax bracket would have to get a yield higher than 5.63% on a tax-free fund when taxable money market funds currently pay 8.4% ($8.4 * (1 - 0.33) = 5.63$).

Not all money market funds are the same. Managers of these funds often increase the maturity of the fund to obtain higher yields. Others load up the fund with riskier commercial papers of companies having lower credit ratings or venture into overseas markets in search of the Holy Grail — higher yields. Extending the maturity in periods when interest rates are coming down makes sense because it allows the fund manager to lock in higher yields for longer times. We would do the same if we know that interest rates are coming down. The trouble is nobody knows with certainty what interest rates will be in the future. We only know the current trend, but we cannot say if this trend will continue in the same direction in the future. Fund managers who try to lock in higher yields by extending the average maturity of a fund could hurt the fund's return when interest rates go up instead of the expected downtrend.

Before buying into a money market fund, an investor must decide on his own, based on his personal risk tolerance level, if the relative small difference in yield justifies the higher risks of longer maturity money market funds. Watch out for funds that have significantly longer maturities than the average money market fund. An investor with a low risk tolerance level should stick with funds that primarily invest in government securities with maturities not much different than the industry average.

Returns on money market funds are stated in seven-day average simple yields and/or daily compounded yields both expressed on an annual basis. A fund with an annualized simple 7-day yield of 8.28%, would have a compounded annual yield of 8.63%. The difference between these two yields has been discussed in the previous chapter. For our purposes, we will use annual compounded yields to evaluate money market funds.

Our portfolio will always contain at least one money market fund because our 5-month emergency money (Axiom III) will remain deposited in a money market fund where withdrawals can be made on short notice without penalty. Since this emergency money in our portfolio will remain fixed at all times, one might ask why not place this money into a CD or other higher yielding investment? The answer is simple and can be expressed in one word: liquidity. We don't know when an emergency arises, but when it does, we have to be able to get our hands on this money on a few day's notice. Money market funds are ideally suited for this purpose whereas CDs are not liquid investments and usually carry a penalty for premature withdrawals.

For example, in the United States money market funds react closely to prevailing Federal fund rates (Fed Fund Rate). Since the average maturity of money market funds is in the range of 30 to 60 days, these funds will follow the Fed fund rate very closely, within a month's time if not closer. Thus, when the Fed fund rate increases, one can expect a similar increase in money market yields. The converse is also true, in other words: Fed fund rates down = money market yields down.

One can never expect spectacular returns from a money market fund. At best they yield perhaps a 1 to 2% premium after tax and inflation. Nevertheless, they are legitimate investment vehicles; they are an absolute requirement in any portfolio structure. Money market funds are a required asset in any portfolio, irrespective of portfolio size, age, and risk tolerance of an investor. It is the first fund a new investor should open and from which a portfolio will be built.

International money market funds are also being offered on the market to investors. These funds invest in short term instruments issued by foreign governments, banks, and companies. They are, however, exposed to currency exchange risks and can vary in return because of changes in currency exchange rates. Again, although their stated return might be higher than what is offered in domestic money market funds, thesc higher returns could be exhausted due to exchange rate changes. International money market funds do not fall into our definition of a money market fund; in our context they are classified as foreign funds.

3.5 Growth Stock Funds

A growth fund, as classified by the Lipper's Gauge, invests in stocks of companies, quote: "whose long term earnings are expected to grow faster than those of the stocks in the major market indexes."

The key word is "expected;" there are no guarantees that these earning expectations will materialize. Growth stocks pay little or no dividends and plow most of their profits back into the company to allow its business to grow with time.

Since equity prices, on a long term basis, are earnings driven, a manager of such a fund has thus the objective to assemble a basket of stocks that, combined, are expected to outperform the market indexes. Few managers can consistently achieve these objectives despite the extensive array of research reports available to them from their analysts. Case in point are the mutual fund performances for the year 1996 (a bull market year) published by Lipper's Guage.

	% Total Return		
	1996	*3-Years*	*5-Years*
S&P 500 stock index	31.6	61.5	152.6
Growth stock funds	25.4	43.6	111.5

This consistent underperformance is a historic trademark of mutual funds and is not likely to change in the future either. Once in a while, a specific fund might outperform the indexes on a quarterly or annual basis, but only a limited few can do it consistently over a long period of time. This doesn't mean we have to exclude growth stock funds from our portfolio. On the contrary, it is a requirement for our own portfolio growth strategy. This is not a critique directed at the mutual fund industry, but a clear indication to us to be very selective in our growth fund searches. We must not forget the above presented performance figures are industry averages; there have been some funds that grossly underperformed and also some that significantly out performed these averages. A mutual fund investor must realize that the fund incurs 0.5–2% annual management and transfer fees and typically holds 5–10% of its holdings in liquid assets. In short, every mutual fund has a built-in underperformance bias which the fund manager tries to neutralize by his "superior" stock picking skills. This is the price we have to pay for having someone else searching out the undervalued and hidden small company stocks that have the promise for significant price appreciations in the future. It is here where a fund manager can shine. Not on a short term basis, because company growth takes time, but on a 3- or 5-year basis. If the fund has delivered above average returns on a long term basis, the fees are justified as long these charges don't deviate too much from other funds in the same class.

In the growth funds selection process, it is important to place the emphasis on long term performances. Don't chase the last quarter's glory performer. Today, Intel, Oracle, Sun Microsystems, and Microsoft are considered classic growth companies. There has been a tremendous technical revolution going on in the computer, software, telecommunication, and electronic fields. The world is rapidly changing and companies participating in this revolutionary trend are the ones managers of growth funds are concentrating on. There are not many growth funds that do not contain any of the current favorites like Intel or Microsoft in their portfolio. But these funds are also loaded with new, unknown upstart companies that might have a new product on the shelf but have not yet produced any profit. Nevertheless, investors and fund managers will gobble up equities of such companies with a vengeance when everybody else is. Prices of these stocks are often driven to unsustainable heights in the hope of having found the new "Holy Grail" of the future. Five or ten years later many of these companies won't be around anymore, having either gone down the tube or, when they have had a successful start, having been taken over by another company. There is never any guarantee that a given growth industry will still be in favor with investors next year or the year thereafter. It could be that sometime in the future someone, somewhere, will invent a product that could make the computer, as we know it today, obsolete. Maybe some medical breakthrough occurs that would make people live beyond 100. Goodbye to today's hot investments, hello new trend, hello new growth company favorites.

Small Cap Funds

Small Cap Funds invest in companies with less than 1 billion in capitalization or total shares outstanding. This is a type of a fund that will tax the expertise of a fund manager to the utmost requiring of him to hold sometimes hundreds of such stocks in the fund. This over-diversification of the portfolio could water down the returns and make his performance look worse than conventional stock funds. His stock picking problems are even more magnified when these Small Cap stocks are in fashion among investors. The strong money inflow in such times might force him to invest part of the new money in mid-cap stocks or force him to close the fund for new investors when he cannot find attractive small company stocks to buy.

Small Cap Stock Funds are not for the faint-hearted or conservative investors because of their high volatility. These funds usually drop more in price than the market indexes during a severe correction (or the start of a bear market). On the positive side, they generally show better returns at the start of a strong upward move (start of a bull market). Of prime importance to an investor in small cap stocks is the question if these type of stocks are currently favored or disliked by the market. Investor's sentiment has a large influence if small cap funds will move up strongly or stay in the doldrums for many months or even years. The Price-to-Earnings (P/E) ratio of Small Cap Stocks or Funds give some kind of indication in this respect. Historically the ratio is higher than their large cap counterparts, but when it is grossly higher or lower than the market itself, it is a signal for us to question our holdings of Small Cap Funds. When investors favor these stocks, they tend to pay a premium price (example, P/E 35 to 70) because of strong expected earnings growth for the next few years. When in disfavor, they hammer Small Cap Stock prices down to levels of valuation that could be below the market index. (For example, P/E 14 when DJ-Industrials is 16). Because of this high volatility, an investor might want to limit holdings of Small Cap Funds to say 10% of his total portfolio assets.

In our strategy we consider holding Small Cap Funds only at the start of a positive bull market move and to switch out of these when the stock market seems grossly overvalued or at the start of a serious correction. Investing in small, emerging companies can be very lucrative and could produce exceptional gains if the fund manager has a knack for picking the future glory companies, the new Xeroxs, IBMs and Microsofts.

Serious investors have always questioned the mutual fund's high management fees while they seem to be unable, as a group on average, to match the performance of an unmanaged basket of stocks that make up the popular market indexes. The industry has lent an ear to these legitimate complaints and now offers index funds.

Index Funds

These are funds that hold exactly all the stocks in a proportion as the index itself. Since no research has to be done by the manager (all he has to do is to match the fund's stock holdings with that of the index itself), management fees will be considerably lower and the fund's ability to match the performance of the index is enhanced. But here again, there will still be management fees and the fund has to hold a small percentage as cash to cover redemptions; there is a built-in underperformance but of much smaller magnitude. Index funds can be of help to investors who believe they are capable of timing the market. At least, when invested, he can expect his fund to perform like the market itself. What a beautiful idea: at last, a fund by which an investor can no longer blame the manager but only himself when things go sour. Only his own skills to make the proper calls to buy and sell such funds will then be the governing factor as to how well his investment performs when compared to the market as a whole. Index funds are considered for our portfolio for this reason because our "Too High-OK-Too Low" concept is built around identifying current excessive over- or undervaluations of the market. In some sense we are thus market timers, but for other reasons than what is generally understood under this term. We only time the market on an infrequent basis and make buy and sell decisions for other reasons than what the typical market timer uses. More on this later.

In bull markets we seek the high returns of Growth Stock Funds; in bear markets we will seek the returns from other investment types or groups, all without ever losing sight of our primary objective of conserving the purchasing power of our portfolio and to let it grow at a steady pace over a period of many years.

Times have changed. Decades ago, an investor could buy a stock of a solid company and store it somewhere for years without having to worry too much about price volatility. These were the times when growth stock investments could be made with a long term outlook of say ten or twenty years. At those times, price volatilty was not such an issue as it is today. Not any more.

Program trading, the arbitrage games played by institutional investment houses and individuals with access to low margin financing, has also increased volatility in the markets. Financial high-wire artists who buy out undervalued and distressed companies with hidden assets, using debt financing to achieve their goals of gaining control of a company, have also contributed to the present volatility of the stock markets. More often than not, these deal makers cannibalize the company by selling off the best assets as soon they gain control of the company. Under the disguise of making the company "lean and mean," such restructuring is often made at the cost of thousands of lost jobs and destroyed careers.

Hedge Funds

Then there are the hedge funds. Let us categorically state at the onset here that these types of funds are not considered for our portfolio because they are purely speculative in nature. Hedge funds take risks the average investor will not and cannot accept. Hedge funds borrow (leverage) from 2 to 15 times the capital they hold to quote, "enhance returns." These are funds very wealthy investors, banks, and investment houses dabble in.

Case in point: consider the billion dollar losses incurred by the hedge fund of *Long-Term Capital* (LTC) in the fall of 1998. Only a massive bail-out (at the urging of the Federal Reserve Bank of New York) to the tune of 3.5 billion dollars could stop the company from facing liquidation. Fifteen major banks and investment houses participated in this bail-out because they were heavy investors in this fund and lost billions. Why the bail-out? After all one might ask, the investors hurt by LTC were not common mortals — they were big players in the markets. Does it signal to us the possibility the Federal Reserve Bank will spring into action whenever the wealthy market players are in trouble, but leave us out in the cold when we, as small investors, are in financial difficulties? The example of LTC just leaves a bad taste.

Not all deals have placed a dark cloud over the investment community. Many mergers and acquisitions have clearly benefited and strengthened the company and made it more competitive in the international market place.

The frequent large changes in market averages, before and during options and futures expiration dates, have clearly thrown the theory of an efficient market out the window. It looks more like the markets are being manipulated by a few powerful individuals and investment houses.

Not only Wall Street but stock markets worldwide, give the impression of resembling a zoo or giant casino during such periods of high volatility. In our investment strategy and technique we will be able to identify periods of extreme buying and selling frenzies and will act accordingly — mostly in a way contrary to what the crowd is doing.

Investors often make the mistake of buying the "hot" fund, i.e., the fund with the best performance in the previous quarter or year. Most growth funds however do not become fully valued for a period of three to four years. Trying to make money by trading frequently in and out of growth funds is most often a losing proposition. In selecting and evaluating growth funds, an investor has to place the emphasis on the fund's long-term performance. Last quarter's star performer could be next year's dog.

In evaluating growth funds, we give major weight toward the 3- and 5-year performance and only minor consideration to its quarterly and last 52-week record. Aggressive growth funds seek out the small emerging companies that have the potential to become the Xeroxs, IBMs or Microsofts of the future. Each of these started out small before they became the mega corporations of today. Because investors and mutual fund managers expect these small companies to grow in the future, growth stock prices will react quite drastically when their earnings don't match investor's expectations. Aggressive growth funds thus exhibit more volatility and risk, facts an investor in these funds must always be aware of. One should not switch frequently in and out of these funds, but, instead should give them the time to reach their full potential, all under the assumption that the fund manager subscribes to the same theory. It wouldn't do an investor any good to stick to a fund while the manager turns over the fund portfolio by 150% each year. Therefore, look in the prospectus of the fund for this turn-over rate.

One should never get emotionally attached to a specific fund. Just because the fund has done well for say three or five years doesn't mean it will keep its performance record. Like it says on the fund's prospectus: past performance is no guarantee for future returns. The advantage of dealing with one mutual fund company for all our fund investments is that it allows us to make a switch from one fund to another with little cost and quite rapidly when our strategy calls for such a switch.

3.6 Growth–Income Funds

Managers of growth–income funds search for stocks that not only have the potential for increased earnings, but also pay out above average dividend income. Certain stock groups such as electric utilities and insurance companies have a record of high dividend distributions. The high income portion of the portfolio tends to offset some of the downside volatility of growth stocks during market corrections. At least that's what market experts proclaim, but it isn't always the case.

There are given market and economic conditions that favor growth–income over growth funds and vice versa. There are also conditions where our strategy calls for holding both or neither of these in our portfolio. This will be discussed in detail later.

The reason we consider both of these fund types for our portfolio is the fact that they are very volatile. In bull markets, the top funds will outperform the market averages, but likewise in market downturns, growth fund losses are usually much larger than the losses in the market averages. However, this is not a rigid fact; there will always be some of these funds that will do better than the market indexes. Investment decisions, when to buy or sell growth funds, must be properly timed. We have mentioned the futility of switching in and out of these funds at frequent intervals because it more often leads to poorer returns than when a simple buy-and-hold strategy would have been used. Neither the frequent switch nor the buy-and-hold strategy will deliver optimum returns.

Our TOO HIGH-OK-TOO LOW strategy reduces the chances of being invested in growth and growth–income funds when the market turns south. In short, we take advantage of the upside volatility of these types of funds. It is well-documented that being invested in growth and growth–income funds during bull market moves, and getting out before the market turns down, is one of the best methods to produce above average returns in a portfolio over the long haul.

Sometimes, balanced and asset allocation funds are being erroneously viewed as growth-income funds because these funds hold equities, bonds and cash equivalents in the portfolio (growth-income funds invest almost exclusively in equities). By investing in asset allocation funds, investors let the fund manager make the call on how to allocate the assets. Since we make our own judgment in structuring our investment portfolio, these funds are not considered in our investment strategy. An asset allocation or balanced fund is structured to make it suitable for a broad spectrum of investors. But there is no optimum asset mix suitable for all investors because each has his own risk tolerance level, different time horizons, and goals. Asset allocation must be designed for each individual's needs and loses its effectiveness when applied to all ages of investors with all different portfolio sizes. Our concept however takes due consideration of these factors.

A multitude of investors in growth-income funds use the dollar cost averaging method in building up their portfolios over the long term. Dollar cost averaging, the passive investment strategy so highly recommended by the mutual fund industry, has serious flaws. From a savings viewpoint, it unquestionably has merits. However, we don't subscribe to the sales argument of investing regularly into stock funds irrespective of market conditions, because equity investments over the long run have given the highest returns. Obtaining more shares when the market is down and less when the market is up doesn't make any sense. This investment style mostly benefits the fund companies because it provides them with a monthly steady inflow of new money and a steady repeat income from the front-end load some of the funds charge. Also, proposals for dollar cost averaging are sometimes used by financial advisors to shield themselves from future customer complaints should they have moved the investors into stock or bond funds at the wrong time.

The belief of being able to buy more shares at a lower price and to sit tight with the ones already owned when the market turns down say 500 points doesn't wash well with investors focusing on potential returns. One doesn't treat a potential terminal disease with a Band-Aid and hope the problem will heal itself in due time. No, on such precarious drops, one must react with major surgery.

Some investors might come into a windfall sum of money due to an inheritance or for other reasons. The question then becomes, should one invest the entire sum all at once or should one spread the investment in equal installments over a period of say twelve or twenty-four months? The possibilities and variations are almost limitless. And so will be the advice from financial experts that an investor will get when he seeks guidance in this matter. One financial advisor might recommend the first alternative, such as placing 33% of the assets in stock, 33% in bonds, and 33% in money market funds all at once (he wants his commission now). Another might advise to go the second route; in other words, the investor is being told to take "advantage" of dollar cost averaging — such as placing a fixed amount each month, say 70% in stock, 20% in bonds, and 10% in money market funds. On and on the possibilities go. We cannot subscribe to either one of these alternatives. Perhaps the best advice would be for the investor to put all the money into a money market fund and to first make himself fully familiar with all the intricate aspects of investing. Then, when the time is opportune, and only in such times, should he plunge himself into the realms of the investment world. Opportune time? By this we mean when the stock, the bond, and/or foreign markets are in a favorable condition to buy as determined by our Fuzzy Logic concept. When the times are unfavorable, no method, be that the "plunge right in" or the dollar cost averaging route will likely be beneficial to the investor.

Last but not least, an investor must be aware that most growth and growth–income funds do not make an attempt to time the market. It pays to read the fine print in the prospectus because most stock funds have the objective to remain fully invested in stocks irrespective if the market is contracting or expanding. An investor in such funds should never expect the fund to be in cash, i.e., having sold the stocks in the fund *before* a severe market correction. No, the decision to be in or out of the stock market is the sole responsibility of the investor himself. Expressed in another way, to protect oneself from losses in stock funds, we as investors have to decide on our own when to buy and when to sell these funds.

3.7 Bond Funds

By investing in bonds, an investor lends his money for a specified time to different governments, government agencies, and/or commercial enterprises. For this, the investor receives an annual interest payment and the promise to get his money back in full at the maturity date. In short, the investor becomes a lender, a banker. Sounds terrific. By investing in a 100,000, 30-year bond at 9% yield, he gets 9,000 per year income for 30 years, and at the end of this period, the full 100,000 back. Terrific. Why not take all the portfolio assets, say 500,000 and pump it into a 30-year bond to get 45,000 every year as income? Enough to go cruising twice a year, play golf once a week, and have a cosy retirement with 500,000 left over at age 95.

Hold it a moment. It's not as terrific as it might look. Just because one has become a lender doesn't give him the right to think like a banker. This line of thinking we cannot afford. Unlike a banker who takes risks with other people's money, it's our own hard earned savings that are at risk here. Having the full faith and credit worthiness of the government or a company to back up the bond doesn't often amount to anything.

Inflation might erode buying power beyond earnings, or the company may have folded.

How could such a massive devaluation ever take place? There might be not enough investors to buy government bonds, the main instruments governments use to finance deficit spending. Thus, a government pushed against this wall might have no other choice than to speed up the money presses. End result after 30 years: hyperinflation with interest rates in triple digits that would make a 500,000, 9% bond almost worthless. A far fetched scenario? Perhaps, but one can never be sure, for it could happen. As bond fund investors we must first and foremost be concerned about the risks involved before we look at the monthly or quarterly dividends the fund pays out. Bonds of higher risk must pay higher quarterly dividends to attract investors.

The question then becomes quite naturally, "does the promised return justify the risk?" Do I want to give a government a loan (i.e., buy government bond funds) when some of this money is used to buy toilet seats at 7000, screwdrivers at 150, and 400,000,000 airplanes? Please be aware that there are no dollar signs behind these figures, it happens in Switzerland (Swiss Francs), Germany (DM) and, yes, also in the U.S. Another question: do I want to buy corporate bond funds that might be classified as junk bonds sold for the purpose to finance a merger, acquisition, or hostile takeover of companies?

Our strategy does not allow for long term commitments to hold bond funds in our portfolio. We invest in bond funds for the sole purpose of taking advantage of prevailing volatilities in interest and inflation rates. In general, when interest rate trends are favorable, i.e., on a downward trend, we buy; if they are stable, we hold; and when the trend is unfavorable, i.e., on an upward trend, we will exit bond funds. Bond funds can be as volatile as equity funds. Bond funds have to be watched with as much attention as equity funds because they too have recently undergone considerable volatility. Many investors got burned in 1994 when the bond market experienced one of the worst contractions in decades. Bond fund purchases and sales have to be properly timed just like stock funds.

Table 3.3 Bond Volatilities

	% Change in Bond Value (yield accounted for)	
Interest Rate Change	*1% down*	*1% up*
6-year Treasury bonds	8	–7
10-year Treasury bonds	11	–9
30-year Treasury bonds	18	–13
20-year zero coupon bonds	29	–18
30-year zero coupon bonds	45	–24

There are possibly more bond funds to choose from than equity funds. There are corporate, government, treasury, municipal, mortgage, and foreign bond funds. Each of these in turn are subdivided into short-term, intermediate, or long-term bond funds, and furthermore can individually vary considerably in risk ratings.

Bond fund yields vary considerably depending on the average maturity and creditworthiness of the debt papers contained in the fund. One must therefore pay attention in the prospectus to these two key factors. The longer the maturity, the higher the yield, but also the higher the risk. Bond funds react inversely in value to prevailing interest rate trends. When interest rates are in an up trend, bond values will fall. Conversely, when the trend in rates is down, bond values will increase. Both of these fundamental attributes in price movements apply to bond sales before maturity. The longer the time span until maturity is reached, the larger the gain/loss effect on the bond's value. As a bond approaches maturity, it will more closely resemble its face value and will exhibit less volatility when interest rates change. This significant factor, important to all bond investments, is demonstrated in Table 3.3 showing the approximate change in bond value as a function of changes in interest rates. The same principle applies to bond fund investments except price changes will not be as volatile because bond funds have staggered the maturities of bonds in the fund portfolio.

Table 3.3 clearly demonstrates the importance that must be given to the average maturity of bond funds before a buy decision is made. It also serves as a reminder that there is no such thing as a risk-less bond. There are short, intermediate, and long term government and corporate bond funds available to choose from.

Since bond fund returns are directly related to interest rate changes, an investor has to keep a close eye on these rate trends. The earliest indication of an impending rate change is given by a change in the Fed Fund Rate, the rate member banks charge each other for overnight loans. The second indication is a change in the Prime Rate, the rate banks charge their best customers. Finally, confirmation of a long term trend change in interest rates is given when the Federal Reserve Bank or Central Bank changes its discount rate.

As shown above, long term bond funds, i.e., funds with maturities of 10 years or longer, are the most volatile when interest rates change. Long term bond funds will only be considered in our strategy when interest rates have been at an unusually historic high level *and* have entered a confirmed downward trend. Such conditions present themselves usually only once or twice in a decade, but these are the times when above average returns can be achieved from bond funds in our investment portfolio. Such interest rate changes from lofty heights will generally also be accompanied by strong positive upward moves in the equity markets. In short, these are the times when our portfolio asset allocation should be heavily weighted in equity and bond funds and very few assets invested in cash and money market funds.

Zero Coupon Bond Funds

Zero coupon bond funds with long term maturities are even more volatile. Zero bonds don't pay interest. In banking jargon, their coupons have been clipped, thus these bonds sell at a steep discount below the maturity value. For example, one can buy a 10,000, 30-year zero coupon bond for approximately 700. At maturity, i.e., 30 years from now, the investor will get the full face value, namely 10,000, paid out to him. These type of bond funds, because of their associated very high risk, will only be considered for our portfolio during interest rate conditions as described above; in short, only when rates are at a historic high and on a definite downward trend. The last time this happened in recent history was in December 1981 when U.S. long bond yields reached 16% and started to come down drastically within two weeks. The downward trend continued for 13 months until the rates stabilized at 9.5% in January 1983. Zero coupon bonds more than doubled during this relative short period of time, a rare occurance for exceptional profits that might not present itself again for decades. This dramatic downturn in interest rates also signaled the start of an unprecedented bull market in equities worldwide. One cannot foresee the future, but it is quite possible that another opportunity for a 4 to 6% drop in long term interest rates will present itself again. When that time comes again, and only then, will we shift part of our portfolio into long term and zero coupon bond funds. We can wait.

Short Term and Intermediate Term Bond Funds

Short term (less than 3-years maturity) and intermediate term (3- to 10-years maturity) bond funds will be included in our portfolio during periods when the yield curve inverted, or in other words, when the 1- to 5-year bonds pay higher interests than long term bonds. Another time to consider these type of bond funds is when yields on money market funds are below our minimum return required, or in other words, when money market fund returns are less than the sum of current inflation and tax rates (plus 1%). Engagements in these short and intermediate term bond funds must be watched closely. When inflation and interest rate are trending up, it would be an immediate exit signal for us to disengage from all bond funds.

Changes in Net Asset Value (NAV) of short term bond funds are never as steep when interest rates change when compared to long term bond funds. In a stable interest rate environment, short term bond funds pay fractional higher yields than money market funds.

Tax-Free Bond Funds

Tax-free bond funds must be rejected for the IRA or other tax-sheltered portions of our portfolio. It makes no sense to invest in a tax-free fund when the investment is already free of income tax. Municipal bond funds should be considered only for those states, cities, and municipalities that have strong prospects for growth. There is nothing wrong in lending a growing municipality money for the purpose of building more and better schools, roads, and other infrastructures. But we do not want to invest in cities that have too many social, financial, and political problems. Especially not in communities where city hall or the state government has lost fiscal control over its expenditures.

For investors in a high tax bracket, municipal bond funds could be an alternative to money market funds in times when the stock markets are contracting (crashes, severe corrections). Instead of parking the assets in money market funds, municipal bond funds might, in such situations, be the better choice since they usually show higher yields (but also more risk).

Mortgage-Backed Security Bond Funds

Mortgage-backed security bond funds will not be considered for our portfolio. Although offering attractive yields, these types of funds don't agree with our investment strategy. When interest rates decline, we expect our bond funds to appreciate in value. Holders of mortgages can refinance their loans to obtain lower rates when interest rates decline. This premature paying-off in mortgages will depress the returns of mortgage-backed funds precisely at a time when we could expect higher returns from bond funds. We do not blame the mortgage holder for trying to obtain lower rates; we would do the same thing. However, when we invest in bond funds, we bank on lower interest rates to give us the extra kick on returns and we do not want the lender to spoil our party.

Current inflation rate trends also exert an influence on bond fund returns. Normally, when inflation is in an upward trend, bond funds returns will go lower. Inflation coming down means bond prices will go up. This is primarily because bond investors bank on the income of the bonds. When inflation heats up, they know the purchasing power of their bond yields will erode over time. In short, the yields on bonds will become unattractive to investors, and they will thus exit from bonds when inflation is in an upward trend. The end result of rising inflation: bond prices will have to go down to make them attractive again to investors, and, since the yield reacts inversely to inflation rates, the yield will go up.

In our strategy, later explained with our Fuzzy Logic approach, we consider holding bond funds when interest and/or inflation rates are steady or in a downward trend. We do not want to hold bond funds (especially not high risk or long term bond funds) when interest and inflation rates are in an upward trend. As simple as that; no ifs and no buts.

3.8 High Income Bond Funds

Moody's and Standard & Poor (S&P 500) are perhaps the best known agencies that rate fixed income instruments (bonds) as to their credit worthiness. Without going into a detailed discussion of the many rating classes, it can be said that the range goes from A to C with Aaa (or AAA) representing the best and Caa (or CCC) the worst rated instruments. Bonds rated B or lower are known in investment jargon as junk bonds. These carry more risk but also pay higher interest.

Financially distressed companies or municipalities with a low rating must pay considerably higher interest to attract investors willing to take on the added risk. It doesn't pay to buy individual junk bonds or to be substantially invested in these types of funds. Junk bond funds have a place in our portfolio provided these funds represent no more than 5% of the portfolio assets as a whole. The question arises, why invest in these risky instruments when, in general, we are against these types of investments? We bank on the fund manager's skills to diversify the fund in such a manner that some of the risks of individual junk bonds are "neutralized" by other, less risky bonds in the fund. It is therefore of great importance for an investor to look at the fund's prospectus with added scrutiny before a commitment is made to buy into the fund. Look for the fund's portfolio diversification, i.e., find out how the fund manager has staggered the bonds as to their ratings. For example, a fund might be made up of the following percentages in each risk class (see table below).

Moody's Rating %		*S&P 500 Rating %*	
Aaa	0.3	AAA	0.6
Ba	3.2	BBB/BB	6.0
B	38.5	B	37.7
Caa	10.8	CCC	1.9
Other	9.3	Other	8.8
Not Rated	37.9	Not Rated	45.0

Weighted Average Maturity: 6.7 Years

Clearly, this represents a high risk fund that might produce an annual return of up to 25% when interest rates decline, but might also lose as much as 10% in as short a period of two month when interest rates increase. We engage in high income bond funds at times of stable and declining interest rates and will switch out of these funds as soon the interest rates start to increase or when the economy is in a downward trend. Slower economic growth can hurt financially distressed companies (high yield bonds issuers) especially hard, more so than large established companies whose bonds are rated B+ or A. Thus, we usually look for an opportunity to sell high-yield bond funds when our economic indicators signal a slowing economy.

Investing in foreign bond funds requires a different evaluation criteria. For this reason, these funds fall into the classification of foreign investments that will be discussed next.

3.9 Foreign Funds

Let us clarify at the onset how we define foreign funds. It doesn't mean we go beyond our own borders and by mutual funds somewhere else. No, by foreign funds we look for funds in our own country that invest overseas.

In 1984, the U.S. stock market accounted for 54% of all the stock traded in the world. Japan was second with 21% and Great Britain third with 7.5%. Total value of the world's equity markets was 2.9 trillion dollars. By 1988, this picture changed dramatically. The U.S. slipped to 25%, Japan increased to 29%, and the U.K. to 24%. Total market valuation worldwide: 11 trillion dollars or, in other words, a fourfold increase in five years. Today, the valuation has changed again; in Japan it contracted and in the U.S., it likely doubled again. As serious investors we cannot ignore the capital markets overseas; they are now all global in scope and closely interwoven with each other. Europeans, Asians, Americans or, for that matter, any other investor group living in a capitalistic country looks across their own borders to find investment opportunities that might deliver higher returns than what they could find in their own country.

During the 1980s, the Tokyo market was the darling of international investors; then followed the emerging markets in Mexico, Chile, Korea, and Taiwan. For a while, the European stock markets were the "in markets" for U.S. investors. But, somehow after all the frenzy, these markets lost favor in the international investment community when investors realized the high market valuations of these markets were purely wishful thinking and represented bubbles ready to burst. Many foreign funds took a bath when the truth and debt of the crisis became fully known. In other words, investors' expectations were not fulfilled and they exited these markets with a vengeance. Nothing will ever go up forever; there have to be corrections to bring exhorbitant markets down to reality again.

There appears to be a real opportunity in the next few decades to make dramatic profits again in the emerging markets in the Far East, Eastern Europe, and South America. These areas might finally recognize that war, political idealism, corruption, or protectionism do not lead to prosperity for the populace as a whole. Economically, they have a lot of catching up to do and it will only be a matter of time before they again will acquire economic power status. Global investing has become attractive because not all markets react in lock-step with each other. When one area market heads south, there will likely be another market somewhere in the world that is booming.

Investors might have banked on these emerging markets because they knew there was a large, cheap labor pool available there with motivated and well-educated workers, especially in Southeast Asia. Some investors had the opportunity to visit these areas in person and came home with the impression that these places had a tremendous economic future. They might have noticed the immense need there to improve the infrastructures and the people's craving for Western goods that would make their lifestyle easier and more comfortable. All that was apparently needed was technological know-how and the capital to get their economy rolling on a fast track. Other investors might not have had the foggiest idea of conditions in these emerging markets but nevertheless invested there because everybody else was or because newspaper reports were full of 40, 50, or higher percent annual profit reports made in these market places. In short, this frenzy represented the herd instinct in its purest form. The speculative bubble had to burst sooner or later. Not only private investors but institutional houses fell prey to these excesses that prevailed at that time. Lucky the ones that got out before the bubble burst — unlucky the ones that got in when the bubble was at its height. Everybody wanted to be in the party, to get on the bandwagon before it was too late.

The same general mentality prevailed among financial advisors and investment houses. They all trumpeted the same tune, namely, one must invest part of the portfolio overseas in order to be properly diversified. We cannot agree with this diversification theory. We are being told that when our own markets take a bath, some other foreign markets might go up and thus cushion our portfolio from a precarious drop. This song is all good and right as long the overseas markets cooperate. They seldom do. Take for example the Swiss Market Index (SMI). This index is made up by more than 80% of the multinationals Nestle, UBS, Credit Suisse, Roche, and Novartis, all of which have extensive business activities in the U.S. Now, if the U.S. market or its economy suffers, so will the SMI because of its large international exposure.

Investors were not wrong in their assessment of the economic potential for Southeast Asian countries. Everybody could see this when one had one eye open toward what was going on outside one's own border. Most, however, failed to recognize that some of these countries overseas were politically not yet ready to participate in western style capitalism. And also didn't realize that excessive market valuation, with P/E ratios higher than 40, sooner or later would have to correct themselves. Then some of these countries had some old fuddy-duddies in high government posts still devoted to the belief capitalism was evil and bad. In others, corruption was rampant among government and public officials, with police forces not being interested in upholding law and order but rather in using brutal methods to suppress the population's cry for democracy and to keep a corrupt government in power.

We do not advocate staying away from foreign investments. No, on the contrary. But we must always remember that things are run differently overseas, economies often operate with other parameters, and what is considered illegal and frowned upon in one's own country is often tolerated. We will invest only in countries that are politically stable, where corruption is not rampant, and the economy is undergoing strong growth. We don't take their leaders' words to make improvements in the area of human rights or empty promises to open their markets to foreign competition. Words are not good enough any more. We want to see action. As investors we can use our financial clout to bring about changes. It's up to them to react or we will withhold our investments from these countries. And that goes for trade partners, too, who want to sell their products in our country but close their borders to our products. There has to be a two-way street. If they don't want it — then no money from us will flow there. It's as simple as that.

Any investor, planning to place some of his money at risk in the foreign market, must consider the associated currency exchange risk. This must be first and foremost in his mind before an overseas investment is being made — even before he decides what type of a fund (equity, bond, or money market) he wants to invest in overseas. Ideally, with foreign investments we search for an opportunity to deliver a double-bonus in returns. First we look toward these markets where equity and bond prices are in a favorable up trend, and, second, where, at the same time, that country's currency appreciates in value against our own currency. In other words, the double-bonus comes when the value of our own currency falls. Under the above scenario, foreign investments have the potential to enhance the overall return of our portfolio.

We do not invest overseas for the purpose of portfolio diversification as many financial advisors advocate. Being invested in mutual funds is enough diversification for us to suit our investment style. Using foreign investments for additional diversification purposes could result in an overkill of this theme and could water down the overall portfolio returns. Besides, many of the mutual funds we invest in already have part of their assets in foreign investments.

Investing overseas by means of mutual funds is perhaps the best and easiest way to participate in foreign markets. We could do it on our own and purchase individual securities through a broker who is versed in dealing with foreign instruments, but this would be cumbersome, costly, time consuming, and carry more risk. We could also, on our own, purchase ADRs, i.e., foreign stocks trading on our own stock exchanges, but here again, we do not have the time to properly research these foreign companies before we invest. Reading financial reports of foreign companies is often an exercise in futility. Some use different accounting methods and some of these reports are often nothing else than a smoke screen, the balance sheets painting a much brighter picture than what it really is. Transparency in financial reports is often nonexistent because different security trading and reporting laws prevail overseas. The old buddy system among insiders is still practiced in many overseas companies. Insider trading is still used in some of these places to the disadvantage of the small investor.

A large mutual fund company that does all the research for us is thus well suited for our purpose. We will only invest in open-end foreign mutual funds. Closed-end foreign country funds trade like individual stocks and are therefore not considered for our portfolio. Such funds can be very volatile, with some of the country funds trading at a substantial premium price to their

NAV. Others, when not popular, will in turn trade at a discount. In short, closed-end funds just don't fit our investment style.

How about opening an account in a foreign bank that offers mutual funds for foreign investors? Forget it. We don't want to open a can of worms when tax filing time rolls around. Investing in our own market is enough of a headache to deal with the local taxing authority. We do not want to make the current time-consuming tax filing any worse; after all, we have a right to enjoy our weekends in the outdoors and shouldn't have to devote additional precious time on such nonconstructive paperwork.

Many foreign countries have their currency pegged against the U.S. dollar. It is therefore of paramount importance to discuss in detail the underlying reasons of the dollar's steady decline in the past thirty years. Very seldom can one read in the papers that a currency's decline is the direct result of governments and people mismanaging their own finances. Instead, there is a lot of finger pointing across the border, accusations being made against foreign countries of not playing fair in matters of international trade and/or setting their own nationalistic interests before global ones.The Swiss complaining about the European Community, Europeans complaining about Americans and — Americans complaining about the Japanese. Always looking across the border to find the culprit; seldom taking a hard look at one's own structural weaknesses. The media doesn't help here either for they, too, participate in intensive "educational" efforts to make people believe all trade and currency problems can be attributed to greedy and unfair practices by foreigners.

We cannot become successful investors if we believe in fairy tales and base our investment decisions on wishful thinking or misinformation. To be successful means dealing with and facing reality, irrespective if the truth hurts our feelings. And, the reality why a currency is losing its value against others can be explained in simple terms.

1. The government and its public have been living beyond its means. It wants the good things in life now without having the money to pay for them.
2. Printing new money and issuing new debt instruments at a rate much higher than the increase in Gross National Product thus transferring its debt to future generations. Let the children and grandchildren pay for the things we want now.
3. The country imports consistently, over many years, more goods and services than it exports — i.e., showing a consistent trade deficit.

4. Investors and the population have lost faith in their own government's ability to keep its fiscal house in order thus taking flight out of their own currency.
5. Government expenditures centered around pork-barrel and social programs enacted with the primary objective to get politicians re-elected irrespective of the harm such programs could do to the financial well-being of the nation.
6. Government foreign policy based on the belief that money thrown overseas can buy friends, votes, and preferential treatment when dealing with foreign governments.

Among many other countries, the United States has been guilty of many of the above mentioned reasons, and then some. The U.S. went on an unprecedented spending spree in the Reagan years to make the U.S. military might second to none. Stationing over a half million soldiers for over 40 years in Europe and Asia cost taxpayers 150 billion dollars a year. While some other countries were concentrating on becoming the world's number one producers of consumer goods, the U.S. placed its priority on being the world's number one producer of space and military hardware.

We do not know what the future holds for our own economy. Large and unprecedented changes are taking place in the world. China is trying to change from a socialistic to a free market economy. Russia was trying to do the same and failed miserably. Japan has its own real estate crisis and has slipped into a recession. Europe can hardly afford its social subsidy programs anymore. Even ultraconservative Switzerland has its own financial and economic crisis because its people don't know if they should join the bureaucratic European Community or stay on an independent neutral course. Who knows what the common Euro currency will do to Europe and the currency markets in the rest of the world after 1999.

In some areas of the world, the economy will dramatically improve; in others, it will decline for one reason or another. Some opportunities will present themselves to make a profit overseas with our investments, but in others it would make no sense to invest because of fundamental social or political problems there. We must decide if we want to invest in the whole region (such as an Asian or European fund) or in a specific country (example, Japan or Germany).

Our personal knowledge of the economic and political situation in given foreign places will govern in which foreign fund we will invest.

3.10 Gold Funds

Finally, a few words about gold fund investments. We are not gold bugs and do not recommend an investor should have a fixed percentage of his portfolio in gold assets as insurance for bad times. Gold doesn't pay interest. Over the past ten years it hasn't significantly increased in value despite the Gulf War, Japan's real estate crash, or the S & L banking crisis in the U.S. Gold today acts in financial markets like any other commodity such as wheat, cattle, and pork bellies. It is not too long ago when the "gold experts" predicted $4000/oz gold prices. Another guru warned of an imminent collapse of paper money and a worldwide financial crisis of never before seen proportions. There are as many crystal ball gazers for future gold price movements as there are for the stock markets. These people are primarily interested in selling their newsletters, out to make profits for themselves, and whose predictions very seldom will benefit the investor who listens to these "forecasts."

Gold however has one important attribute. Throughout history, one ounce of gold has bought a man's suit. In Roman times, the Middle Ages, and even today, gold has not lost its purchasing power unlike paper money in any denomination. There could come a time when the U.S. dollar's current status in the world as the preferred currency would be lost and vast holdings of dollars worldwide would be dumped in panic-like fashion causing a severe devaluation of this currency. Worse, it would trigger a worldwide financial catastrophe. But, since we are not soothsayers, we handle such an eventuality when it occurs, not before. This then would be the time when gold assets could become the preferred investments as they were in the Middle Ages and earlier. That will be the time to react and to change our investment strategy. Not before. For now, gold and gold fund investments are not being considered for our portfolio. The same goes for gold jewelry and coin collections. Jewelry is nice to wear, coin collections a nice hobby, but both are not what we consider viable investment instruments.

We will move into gold investments when the remote possibility of a worldwide financial crisis erupts and then we will consider only gold mining funds. Such a possible worldwide crisis will not happen in just one day. It will creep up slowly, will move from one country to the next, from one geographical area to to next, and we will have plenty of opportunity to adjust our portfolio if needed. Should these times arrive, and we all hope they never will, we can also observe the gold price to make gigantic upward movements in price, not 10 or 15% but in the magnitude of 20 and 40% in as short a time as six months. It is then when the crisis is serious and we have to think about protecting our nest egg, not before.

3.11 To Hold or to Sell Funds?

Many investors have become so obsessed with their investments that they have to rise every morning in the wee hours and head straight out to get their hands on the latest newspaper. The first thing they look for is the business section to find out how many cents their mutual funds have changed on the previous day's trading. Within this group of investors we will find those who trade in and out of funds at frequent intervals, falling prey to the daily or weekly volatility of the marketplace.

Table 3.4 Gains Needed to Break Even After a Loss

Percent Initial Loss	*Percent Gain Needed after Loss*	*Percent Initial Loss*	*Percent Gain Needed after Loss*
10	11.1	50	100
15	16.4	55	122
20	25.0	60	150
25	33.3	65	186
30	42.9	70	233
35	53.8	75	300
40	66.7	80	400
45	81.8	85	567

In the other extreme, there are also those who will not check on their fund performances for many months, usually waiting for this information until they receive the quarterly report from the fund company. These are the passive investors who more likely will dismiss large losses in NAV with the faulty attitude that it is only a paper loss, believing that the drop had nothing to do with the fundamental "strength" of the underlying securities in the fund's portfolio. These investors believe the markets will heal themselves in due time. Unfortunately, due time might drag itself out for years. We cannot afford to manage our assets with such knee jerk reactions or passive attitudes. We don't want to play ignorant to the fact that financial markets have changed and fresh approaches are needed to master the rapid changes that have become a common occurence since electronic program trading has been introduced. A loss is a loss, irrespective if the asset is held for the long or the short term. There is just no better way to explain what an investor has to face up to after a large loss in a fund's NAV than what is shown in Table 3.4. Passive investors often overlook the fact that an investment, having lost 50% of its value, will have to gain 100% just to return to the level it was before.

We give this reality due consideration in our investment strategy. If an aggressively managed fund dropped say 38%, we would need a gain of 61% to just break even again. We don't hold on to an underperforming fund in the hope it will come back. This might require a waiting time of several years; years of lost opportunities.

After a severe market drop or when fund companies report large losses at year's end, it is not uncommon to hear fund managers tell their investors the stocks in his fund are sound long term holdings and that investors have unjustly driven the markets down because of fear and panic. In short, they recommend their clients not to sell, to hold on, that the stocks in the fund will come back and represent good value. Naturally. The fund manager will not admit he loaded the fund with turkeys or used the wrong investment strategy. Instead he usually places the blame for the losses on others, never on himself.

We attempt to be invested during market up-moves and out of a given market on down-moves. We also recognize the fact that our concept will likely not make it possible for us to get in at the exact bottom or out at the top of significant market moves, but the TOO HIGH-OK-TOO LOW concept will tell us when it is or is not appropriate to participate in a market. We also don't want to become slaves to our money and forget that there are other things in life important to us to pay attention to like our spouses and children.

3.12 Keeping Informed About Fund News

The larger our portfolio grows, the more time we will have to devote to keep up with the latest financial news. Situations can rapidly change as many investors had to learn the hard way in the week before the October 1987 crash. For some it might be a dreaded chore to devote time to financial matters; for others it might be an exciting hobby. And, unfortunately for some it is *the* only thing in life that keeps them interested.

Unquestionably, a computer is a great asset to have to do these chores and to implement our investment strategy. But, it is not an absolute requirement. Our concept can be followed without a computer although it would be more time consuming. Throughout this book we will present samples of spreadsheets with appropriate formulas to help the reader structure the investment program to his own specific needs.

Most public libraries carry the latest issues of newsletters, magazines, and books related to the subject of investing. These provide the learning materials for individuals to become proficient in this field. Not every investment recommendation is foolproof, but it helps the reader to become better informed and make better investment decisions. By reading a lot of these publications, one invariably will be exposed to many divergent opinions about how best to invest money and in what direction the markets will be moving. This is not a bad thing at all; opposing viewpoints are the life blood of the markets. Being exposed to these different opinions provides us with the learning curve we need to become successful investors.

If we have access to the Internet we should also not forget to browse the vast amount of information available in the Investor Forums periodically and the homepages of well-known publications, investment houses, and fund advisory services. In this medium we can download just about any historical performance charts of mutual funds of interest to us. There are history graphics for the stock market, interest rates, economic data, and just about anything the reader desires. And, most of this information can be obtained free, with only the telephone charges from our own pockets. But, here again, there will be some garbage in the pile. After a little experience surfing the Internet, the reader will quickly be able to separate the downloaded information that is of value to him and that which is of no value.

With the publication of this book, the author also maintains a homepage on the Internet where the reader can view the current Fuzzy Logic conditions, risk, and asset allocation tables and much more — information that saves the reader the time to do the research on his own.

3.13 Investment Advice Pitfalls

We must be on guard and categorically reject any investment recommendation that

- Is made over the phone by a boiler room operator (telemarketing) promising short term profits of 15% or more,
- Is based on brokers' and advisors' projections at what time *and* at what level the market will be in the future,

- Is based on *one* single technical, fundamental, or market indicator,
- Is based on any astrological or voodoo indicator,
- Is based on a hot tip or alleged insider information,
- Involves an investment we are not familiar with,
- Involves an investment whose risks are beyond our risk tolerance level or whose risks are not known to us,
- Involves an investment that doesnt meet our portfolio risk limits.

3.14 Fund Turn-Over Ratios

We place a fund's turn-over ratio under close scrutiny. Funds with high turn-over ratios (more than 70%) pay transaction fees to the brokerage houses when they either buy or sell assets. These fees come directly out of the fund's assets and can depress the overall return of the fund. Funds with a turn-over ratio of 130% or more must be viewed as playing the short term trading game, i.e., are considered in our viewpoint as "market timer" funds. Keep in mind that a turn-over ratio of say 200% means the fund's assets are sold at an average of once every 6 months. Likewise, a fund with 33% turn-over means it's holding the assets an average of 3 years. And we all know how adept fund managers are at timing the market. Some are good timers some of the time, but very rarely good all the time. Case in point: take an emerging growth fund with a turn-over ratio of 180%. Now, we all know for emerging companies it takes time to mature and to become recognized by investors. As investors, we would thus hold such an emerging stock for the long term to give it the time it needs to mature, say 3 to 10 years. What the above example indicates is that the fund manager is not very adept in finding these emerging nuggets in the market; otherwise, he would not turn over his portfolio at a rate of once every 6 to 7 months. In short, he is shooting craps in the investment casino. And for this, he gets paid an annual salary in the six figures. Some manager, some exuberant overrating.

We prefer funds that have a turn-over ratio of less than 50%. That means the average holding period of a fund asset is 2 years or more. Don't let the fund manager gamble with your money; use your own strategy to decide when to buy and sell.

Some mutual fund companies, known collectively as institutional investors, have become so large that they have the power with their buy and sell orders to move a given equity in a significant way either up or down. We venture to state they also have the power to move the stock indexes as a whole under given market conditions. Strong money inflows into mutual funds normally generate upward momentum to the market indexes. Conversely, panic selling by mutual fund investors can force the large fund companies to sell substantial portions of their holdings and thus contribute to the downward trend of a market. This effect will become more pronounced as more and more private individuals invest their savings, IRAs, and pension money in mutual funds instead of bonds, savings accounts, and CDs. The 1990s have delivered an unprecedented bull market in equities and a large part of this strong upward momentum has been the result of new mutual fund money inflow. We only hope the investors that got into this market in the past few years realize that the good times will not roll on forever. Twenty and thirty percent annual returns on funds cannot be expected to continue for indefinite times. Buying on the dip, when the market has gone down 5 to 10% has not been a bad investment technique during this raging bull market. But, sometime in the future, when we don't know, buying on the dip might not be a good idea anymore because more dips could follow each other in short succession. In other words, the markets will go south and signal a major bear trend. How long will the mutual fund investors tolerate negative returns in their portfolios until they throw in the towel and move away from their long-term holding strategy? We don't know. The only thing we know is that when this happens, i.e., massive liquidation of mutual fund holdings, the good times are over for investors, mutual fund companies, and fund managers. With our investment concept we expect to be out of equities before this happens because we will generally not buy and hold substantial portions of our portfolio in equity funds when the market is overpriced, i.e., shows excessively high P/E ratios.

4 Selection of Mutual Funds

To start, we collect information and performance data for various funds of interest to us. This information can be obtained from the financial papers. We first search for fund companies that offer a great number of funds in all the groups of investments we consider. Information on funds can also be obtained from the fund company itself, its homepage on the Internet, or by subscribing to a mutual fund newsletter. Don't be shocked by the legal language of fund supplied prospectuses. These are written by lawyers for the benefit of lawyers. Just look for the important nuggets.

4.1 Funds to Watch

From brochures, newsletters, and financial papers we compile a list of funds that will become the core funds we will monitor each month. The most important requirements for funds to be included in our list are:

1. The fund must have been in existence at least 5 years, i.e., 3- and 5-year performance data must be available,
2. Front loads not higher than 2.0% and no back-end loads if fund held for longer than 3 months,
3. No hidden 12-b charges (we prefer no-load funds),

4. If investment restrictions apply: minimum initial investment required should not be higher than $5000 and minimum repeat investment thereafter: not higher than $250,
5. Select only fund companies that provide for telephone or electronic fund exchange privileges,
6. Preferably invest all funds with one mutual fund company instead of spreading fund holdings over too many investment houses.

4.2 Funds to Consider

4.2.1 Money Market Funds

Search for funds with maturities nearly equal to the average typical money market fund (MM) of 40 to 75 days. Do not consider MM funds that invest substantially in foreign instruments. Consider only MM funds that offer check writing privileges.

4.2.2 Aggressive Growth Funds

Look for growth funds that remain fully invested in equities irrespective of market conditions. Do not include balanced or asset allocation funds that switch assets between equity, bond, and cash holdings according to their own investment strategy. We ourselves will make the decision how our portfolio assets are to be allocated.

4.2.3 Growth–Income Funds

Suitable funds are those that contain a large portion of high dividend paying equities. Not suitable are funds that derive the major portion of income from the fund's bond holdings. Included should also be a so-called market index fund that attempts to mirror one of the equity market indexes.

4.2.4 Intermediate and Long Term Bond Funds

Consider funds that invest in government, treasury, and/or corporate bonds with maturities of 5 to 15 years. The major portion of bonds in the funds should be rated "A" or higher.

4.2.5 High Income Bond Funds

Look for funds whose historic returns in bull bond markets have been significantly higher than the average intermediate bond fund in the same period. The majority of bonds contained in these funds should be rated "B." Include also bond funds that contain a large proportion of zero bonds in their portfolio.

4.2.6 Foreign Funds

Include only funds that invest in the geographic areas whose political, financial, and economical situations are known to you. Also include in this watch list one or two funds whose investments are not restricted to a specific geographic area or continent, i.e., funds that invest in all foreign markets where, according to the fund manager, opportunities for gains present themselves.

4.3 Keeping Track of Fund Returns

Fund performances are reported in different ways by fund companies. Some report how a 10,000 investment has grown over a certain period of years. Example: 10,000 invested on Dec. 31, 1992 is now worth on Dec. 31, 1997: 23,000. This is a popular way for fund companies to report how their fund has performed. Some even supply a nice graphic to show this growth pattern. Many overseas fund companies play games with these data. For example, when a fund had spectacular returns in the past four years but a dismal performance five years ago, they conveniently leave out the fifth year in the graphic. The best way to show this data, in our opinion, is when the graphics also show how the fund performed as measured against a benchmark index. Now this would be really valuable information for investors, but very few fund companies do this. They are more interested in putting their fund perfomance in the best light possible even if this means leaving out some pertinent data.

Others would state this as a total return of 130% over the past 5 years. For our purpose of fund evaluations, we use average annual total returns. Total returns can be converted by using either Table 4.1 (3 years) or Table 4.2 (5 years). These tables have been developed to eliminate the compounding effect of an investment when held for a prolonged period of time. The compounding effect, as we have discussed in Chapter 2, has nothing to do with a fund manager's skills to pick the right assets for the fund. With the

Table 4.1 10,000 Invested after 3 Years

Value after 3 Years	Percent Total Return	Average Annual Return	Value after 3 Years	Percent Total Return	Average Annual Return	Value after 3 Years	Percent Total Return	Average Annual Return
25000	150	35.7	16000	60	17.0	13000	30	9.1
24000	140	33.9	15500	55	15.7	12750	27.5	8.4
23000	130	32.0	15000	50	14.5	12500	25	7.7
22000	120	30.1	14500	45	13.2	12250	22.5	7.0
21000	110	28.1	14250	42.5	12.5	12000	20	6.3
20000	100	26.0	14000	40	11.9	11750	17.5	5.5
19000	90	23.9	13750	37.5	11.2	11500	15	4.8
18000	80	21.6	13500	35	10.5	11250	12.5	4.0
17000	70	19.3	13250	32.5	9.8	11000	10	3.2

Table 4.2 10,000 Invested after 5 Years

Value after 5 Years	Percent Total Return	Average Annual Return	Value after 5 Years	Percent Total Return	Average Annual Return	Value after 5 Years	Percent Total Return	Average Annual Return
35000	250	28.5	26000	160	21.1	17000	70	11.2
34000	240	27.7	25000	150	20.1	16000	60	9.9
33000	230	27.0	24000	140	19.1	15000	50	8.4
32000	220	26.2	23000	130	18.1	14500	45	7.7
31000	240	25.4	22000	120	17.1	14000	40	7.0
30000	200	24.6	21000	110	16.0	13500	35	6.2
29000	190	23.7	20000	100	14.9	13000	30	5.4
28000	180	22.9	19000	90	13.7	12500	25	4.6
27000	170	22.0	18000	80	12.5	12000	20	3.7

average annual total return we have a yardstick that is truly indicative of a fund's performance. In our example above, the fund would have had a corresponding 18.1% average annual return.

We perform this analysis of fund returns at least twice a year, or each time before we make a major adjustment in our investment portfolio. We narrow down the number of funds for consideration to 20 and list them as shown in Table 4.3. The best and worst year performances are obtained from mutual fund newsletters or financial papers that periodically report monthly or quarterly fund returns.

Table 4.3 Mutual Funds Performance Index

Fund Name	*Fund Group*	*Average Annual Total Return*		*Worst Year*	*Best Year*	*Empirical Performance Index*
		3 Years	*5 Years*			
D	Aggressive Growth	19.5	17.8	–1.9	54.9	13.9
C	Aggressive Growth	20.3	16.7	–2.5	48.3	12.7
B	Aggressive Growth	18.9	17.7	–6.1	46.3	11.2
A	Aggressive Growth	19.6	16.9	–4.5	38.9	10.9
F	Equity-Income	14.3	13.2	1.0	31.6	10.3
E	Equity-Income	16.2	15.1	–6.8	41.8	9.2
G	Equity-Income	13.4	12.8	–6.4	28.7	6.8
H	Equity-Income	16.4	13.9	–14.1	29.4	4.6
I	Intm./Long Bond	5.9	7.5	–2.1	20.6	7.9
K	Intm./Long Bond	6.5	8.3	–5.4	21.1	7.5
L	Intm./Long Bond	6.2	8.1	–5.2	17.7	6.0
M	Intm./Long Bond	7.5	9.4	–12.3	20.8	6.3
N	High Income Bond	13.1	14.2	–1.8	34.3	14.9
O	High Income Bond	12.2	13.7	–4.6	29.8	12.7
S	Foreign	12.6	7.7	–6.6	69.3	8.9
Q	Foreign	13.2	8.6	–4.5	32.3	6.4
P	Foreign	15.6	10.3	–11.2	39.5	5.4
R	Foreign	13.6	8.3	–27.2	63.9	0.5

The minimum return needed is calculated by the previously discussed formula,

$$\text{minimum return needed} = \frac{1 + \text{inflation rate}}{(1 - \text{tax rate})}$$

The key data in Table 4.3 are the performances of each fund in the *best* and *worst* years, variables that give us an insight into a fund's potential volatility without having to use mathematical statistics.

For each performance variable we assign a certain weight that reflects the importance we place on these entities. For example, to equity funds we place a 15% weight on the 3-year and a 35% weight on the 5-year performance. Likewise, the worst year receives a 40% and the best year a 10% weight. Bond funds have different weights assigned.

4.4 Performance Index for all Equity and Foreign Funds

$$\text{Performance Index} = 0.15\ a + 0.35\ b + 0.40\ c + 0.10\ d \qquad 4.4.1$$

4.5 Performance Index for all Bond Funds

$$\text{Performance Index} = 0.30\ a + 0.20\ b + 0.25\ c + 0.25\ d \qquad 4.5.1$$

where a = 3-year average annual total return
b = 5-year average annual total return
c = total return in worst year
d = total return in best year

It is important to mention that the Performance Index is not an indicator for future returns. Its sole function for us is to screen out those funds that, by our criteria, showed the best performance within the group of like funds. With this method we attempt to weed out the short term "glory" performers and prevent us from following the crowd that normally invest in last year's or last quarter's high flier fund. Very seldom will such high flying funds be able to repeat their previous short term performance. Buying into such funds can, more often than not, be a disappointment. We must again emphasize that our strategy centers around holding a fund for the long term, say 5 years, provided it keeps on delivering returns that meet our requirements.

Also be on guard against selecting funds that have shown exceptional and much higher returns in the past 2 years than like funds in the same group. These turbo-performers might be loaded with highly risky stocks, bonds, or even enhanced their performances by means of speculative derivative holdings.

The data presented in Table 4.3 are actual performances of funds within one large mutual fund company. Names of the funds have been replaced by letters for obvious reasons. Besides, it is not our purpose here to point fingers at individual high and low flying funds; we only want to demonstrate with these examples how our fund screening method works. This is not an absolutely reliable method because other unknown factors such as changes in fund managers or investment objectives might result in completely different performances in the future.

4.6 Fund Managers

The hypercharged bull market of the past few years has brought forth a number of very young fund managers who have never experienced a protracted bear market. Many of these high paid managers were still in business school during the 1987 crash. Some don't know how destructive a 25 or 40% drop in the stock market can be on private investor portfolios. The financial market crisis in the Far Eastern, Russia, and South American regions have not yet run their full course . At this time we don't know if the crisis will encompass the entire globe or if governments finally come up with lasting solutions to douse the fire and make some real changes to prevent such financial calamities from recurring in the future. Only time will tell.

Looking at fund prospectuses one often comes to the conclusion there are managers primarily interested in riding the bandwagon; they don't want to lag their peers in performance, and thus might expose their fund to risks they should not take. This bull market cannot roll on to eternity; there have to be corrections to weed out the excesses sooner or later. That's a factor many of these young managers don't want to consider — let the good times roll — managers not even 35 years old and already handling billions in fund assets and drawing salaries close to half a million a year. We venture to say that many of these young managers won't be around once this bull market has run its course. They might be forced to retire at age 40. The ones who will survive because of their superior performance will likely leave the fund later and open their own investment company.

Be on guard when a fund announces a switch in managers. Funds that are managed by new people have to be more closely watched than others because these new managers might bring a new investment style into the fund. It doesn't mean we should exit this fund when a manager change is made. We should by all means give this new guy a chance to show his stuff. But, if he doesn't live up to our performance standards after one or two years, we dump his fund. It's our prerogative to do so; after all, it's our money, not his.

Take the time to study the fund prospectus before investing in a new fund. Many of the funds are not what they proclaim to be. A fund might be advertised as a domestic equity fund, but upon closer examination, we might find out that it holds up to 30% in foreign equities.

Also, as mentioned earlier, keep a close eye on the management fees. Avoid funds whose fees are significantly higher than like funds in the same group.

In some cases the charges can be so inordinately high that the total return of the fund doesn't even cover the current rate of inflation. Also avoid funds that show high redemptions, i.e., funds that have fallen in disfavor with investors. High rates of redemptions might force the fund to sell assets at an inopportune time, times when equities or bonds are in disfavor with investors and thus in a downward trend.

It would be wrong to invest only in those funds with the highest Performance Index and ignore other important factors such as the associated risks involved and the effects such an investment would have on the overall goals for our portfolio asset allocation, subjects that will be discussed in later chapters.

Many investors fall prey to advertisement hype and rush out to purchase the best performing funds of the past quarter or 12 months. More often than not, these high flyers could take a severe tumble when the stock market heads south. Furthermore, the percent drop in value for such funds will likely be more severe than the market as a whole. What we are looking for are not funds with short term glory performances but "perma-burners" that produced consistent high returns in their group of peers over several years.

We also prefer funds where the manager has successfully run the fund for several years. He should have had the opportunity to show his expertise in fund management as a fund manager or assistant for at least five years.

4.7 Where to Buy Funds

We don't place too much trust in bank tellers' and stock brokers' recommendations for fund purchases. The danger of being sold a "turkey fund" cannot be overlooked. We prefer to deal with the fund company directly, not through a middle man. Large banks and brokerage houses run their own mutual funds and can thus offer investors a wide variety of investment possibilities. They are not better nor worse in performance than the funds sold by mutual fund companies. But, funds are only another branch of what they can offer to the public. They also sell individual stocks, CDs, Treasury papers, options, futures, you name it. They may even be more interested in providing service to investors with large portfolios and don't pay as much attention to us mortals with just a few thousand to invest. Besides, most of these funds carry a front-end load of up to 8%, a commission to the benefit of the bank or brokerage house which we can do without.

Be very skeptical of fancy charts they usually present to show how "well" their recommended fund performed in the past five years. The fact is, the last five years (1993–1998) have been unprecedented for stock investments and it is therefore not difficult to show a good performance. As a close friend of mine once said: "Every idiot can make money in a raging bull market. True investment skills show up when the market goes down." Let the salesperson show you performance data for the given funds in the year 1990 and 1987; that's where the expertise of a manager came truly to the forefront. And let us never overlook the meaning behind the standard phrase: *past performance is no guaranty for future returns.* No reputable salesperson of mutual funds will ever make an attempt to sell you a fund with a promise of x-percent returns in the future based on past performance.

In contrast to banks and brokerage houses, a mutual fund company will take on all comers — the rich widow, the small saver, large and small accounts, it doesn't matter too much. They specialize in these products and their salespersons usually know the fund business inside and out. This is why we prefer to deal directly with mutual fund companies.

Recently, an elderly Swiss woman inherited over 100,000 Swiss francs. Not being familiar with investment possibilities, she went to one of the three big Swiss banks for advice on how to best invest this inheritance. She specifically told the investment advisor to invest the money conservatively to protect her principal. All she was looking for was a 4 or 5% return because savings accounts paid at that time a measly annual 3%. After all, she was already 83 years old. What was the advisor's answer? — No problem, the "advisor" proclaimed, he had just the right things for her. Mutual funds, diversified to reduce the risk. 60% in an Asian emerging market fund and 40% in an Ecu-bond fund. Was she impressed. Imagine, her being invited into a lavish room of the bank to discuss investment manners. And how impressive and knowledgeable the advisor seemed to be. Clearly, to her this advisor knew this business of investing inside and out. Therefore she agreed to let him invest her money as he recommended. Six months later, her portfolio lost 13% of its value. Some diversification, some "expert" advice, some conservative investment.

What happened? We don't know. Most likely the "advisor" was a novice and just followed his superior's advice to get people into the then hot Southeast Asia and Ecu-bond funds. Or worse, he might have been promised a special commission if he could "unload" these funds on the public. — Need we say more?

5 Risk

We now discuss the most important and controversial subject related to investments, namely, the question of how much risk an investor should take and how one can control these risks to avoid poor portfolio performances. Always present and always in dispute is the difficult question how risk should be measured in a meaningful way and applied in investment decisions. To this day, no clear cut definition of risk is in existence because risk has different meanings among different investors. No one has been able to clearly assign a risk surrogate that could stand up universally to any type of investment.

The widely followed methods used by the financial community and academia to define risk involve elaborate calculations, hence most fund investors either overlook risk analysis entirely by focusing on returns only, or rely on so-called experts to guide them toward the proper type of investment that would suit their investment style. Clearly, this is one of the most serious errors an investor could make. We agree with these experts on one thing, an investment program can only be effective when it includes some form of risk control. However, we completely disagree with the methods the experts use to define risk.

5.1 Diversification

At the forefront of risk control is the widely held belief and much publicized theory that an investor must diversify his portfolio among several types of investments and also among several individual holdings within an investment type itself. So popular has this theory become that the mutual fund industry uses diversification as one of their strongest sales arguments for pushing their funds on the small investor.

Unfortunately, diversification has become a much abused buzz word and is often touted as the all-encompassing solution to the problem of risk management. Don't place all your eggs into the same basket, they say. Then, this theory could compel us to believe that if by a stroke of bad luck, there happens to be a loser in the basket, it wouldn't affect our portfolio too much because it would be neutralized by all the other winners in the basket. How convenient. Wouldn't it be better to get rid of the bad investment that proved to be a loser and replace it with one that has the potential for better performance in the future? Holding on to a loser in the hope that it will recover some time in the future is not compatible with our investment strategy. In other words, diversification isn't the ultimate solution as many financial advisors try to make us believe.

By now, the reader must have guessed that we don't place very much value on diversification as a strict and unbendable rule when we make investment decisions. We advocate placing our investments into those things that, at present, represent an opportunity to produce a return at reasonable risk, consistent with our investment objectives. If there are opportunities in two or three different types of investments to reach these goals, we would invest in all of them. But, there can also be times when only one investment type meets our criteria, in which case we would not hesitate to place all our eggs into this type alone. For example, when present equity, bond, or foreign fund conditions are unfavorable, we wouldn't be adverse to placing all our funds into a money market fund until such time when conditions would improve again for the mentioned fund types. We don't want to become bogged down in our investment decisions by a single and rigid diversification rule.

5.2 Derivatives

Another popular method used by professionals is to invest in derivatives to quote, "reduce risk." Derivatives are financial instruments supposed to provide insurance against large losses. For example, a fund manager might hedge the foreign assets in his portfolio by buying forward currency contracts. In doing so, he attempts to reduce the fund's exposure to foreign currency risks. In other words, the fund manager believes that the currency of that country he is invested in could be overvalued or undergo depreciation in the near future. By hedging his portfolio against a possible devaluation of that currency, he hopes to protect some of the gains he has made with these foreign investments.

The problem is that this kind of "insurance" is expensive. When the fund manager's intuition of a currency fall doesn't pan out, the costs for this hedging come directly out of the fund's assets and thus lowers the performance. This is only one example; there are many other derivatives used by fund managers to either reduce risk or enhance the performance of a portfolio.

We don't exaggerate when we state derivatives are mostly profitable for the ones who create and sell them. The small investor, more often than not, ends up holding the short end of the stick. There are options, futures, straddles, saddles, and whatever else the financial community might dream of next to make money primarily for brokers, bankers, and investment houses, but not necessarily for the investors. We venture to state that bankers and fund managers wouldn't commit themselves to such massive derivative trading if they would fully understand the true risks of these trading practices.

Lately, the financial papers have reported bank failures and investors losing millions on deals with derivatives. But, despite many failures, speculators and investment houses will dream up new and more exotic derivatives in the future to keep this speculative bubble going. Why? Because there always have been and there always will be people who believe in the fairy tale of making money without risks.

We view derivatives as just another type of highly speculative venture and refrain from investing in any fund that uses derivatives extensively to hedge its portfolio assets. Therefore, investments in hedge funds are not considered in our investment strategy.

5.3 The Beta Factor

Perhaps the most widely accepted measure of risk is the so-called Beta of either an individual security or a portfolio. Specifically, Beta expresses the relationship between a portfolio's or security's return against the return of a benchmark index as a whole. Beta is also defined as a measure of volatility. A fund with a Beta higher than 1.0 is thus more volatile and a fund with a Beta below 1.0 is less volatile than the benchmark market index. In the following formula "Std. Dev." stands for "Standard Deviation."

$$\text{Beta} = \frac{\text{Std. dev. of the fund or portfolio}}{\text{Std. dev. of the benchmark index}} \qquad 5.3.1$$

Let's use two examples to demonstrate what Beta means and what kind of thought process it generates in an investor's mind.

	Return	*Beta*
S&P 500 Market Index	15.0%	1.000
Fund A	20.0%	1.333
Fund B	10.0%	0.666

According to the Beta theory, Fund A will be 1.333 times riskier than the market and Fund B with a Beta of 0.666 is the less risky one of the two.

One can guess the kind of sales talks these two funds might generate:

Mr. Bull, Broker extraordinaire: "We are in the greatest bull market in history. The Market will go to 15,000. Buy Fund "A" because it has outperformed the S&P 500 and thus has the potential to deliver fantastic returns in the future. Forget the 8% load; the fund will compensate for it many times over by its consistent higher returns year after year."

Mr. Bear, Broker extra-contrary: "The market's had it; it's going south. Market breath is bad. We have just formed the right shoulder in the classic head-and-shoulder pattern. Tell you, it's time to shift gears. Sell Fund "A" and buy Fund "B" which is less risky because it is less volatile than the market itself. Besides, the fund is loaded with recession resistant stocks. Fund "B" is the kind of protection one needs in times of uncertain market conditions."

Mr. Armadillo, Certified Financial Advisor: "With clear evidence at hand that everything about the market is now unclear, it's time to take a defensive posture. Diversify. Balance the risky Fund "A" with an equal amount of Fund "B." Also raise the cash in the portfolio from 5 to 20%. This way the portfolio is protected on the down-side but maintains its upside potential when the market keeps on going up."

Finally, the investor had enough of this conflicting sales talk: "Just a moment.... Hold it... Gentlemen. Please explain to me in straight and clear terms how much my portfolio value will drop when the stock market should go down say 20%?"

Certainly they can. An investor advisor's jar is full of all kinds of cookies (excuses). They will tell us that an investment risk cannot be defined by its

Beta alone. There is also an Alpha plus a constant error factor that has to be factored into risk analysis. Out they come with the cookie that states

$$\text{Total Risk} = \text{Alpha} + \text{Beta} * (\text{Market Return}) + \text{Error Factor} \quad 5.3.2$$

This is called the Total Return equation which states that an investment's return is governed by a constant Alpha (accounting for factors not related to the market and thus diversifiable), the Beta (accounting for market moves and non-diversifiable) and an error factor (that accounts for all the other factors that cannot be explained by either the Alpha or the Beta). How convenient. If the investment behaves contrary to the Alpha and Beta factors, then the Error Factor is usually adjusted to fit the event. These advisors never mention that the Error Factor includes the miscalculation on their part in having sold us a turkey investment in the first place.

5.4 The R^2 Factor

Then investment advisors also use the so-called R-squared factor to explain why an investment didn't perform as expected either to the up or down side. R-squared measures the correlation between an investment and the market as a whole and ranges between a value of 1.0 and zero. A fund with a R-square value of 1.0 correlates exactly to the market; a value close to zero indicates that the fund doesn't correlate to the market. For equity fund managers who deal in individual securities, these factors might be useful yardsticks. However, for the average individual investor, when his mutual fund lost money, Alpha, Beta, and assorted other factors don't mean a thing.

We have a hard time accepting the theory of controlling the Alphas and Betas to manage the volatility of a portfolio. The fact is that these factors are based on past performances and make the assumption that the future is just an extention of the past. In short, this theory is founded on the misguided belief that one can predict the future based on the past.

Worse, using one Beta both for up as well as down markets has its own pitfalls. There is no plausible evidence that a fund will react with the same volatility in up as in down markets when measured against the market volatility. The exception is the pure index fund that contains in its portfolio exactly the same proportion of securities as the index upon which the Beta is used for comparison. Most equity funds however exhibit two different

Betas, one for up and another for the down market. To complicate matters further, these Betas don't remain the same for extended periods of time. They change frequently because the fund manager usually makes portfolio adjustments, if not weekly, then certainly monthly. In other words, by the time an investor reads about a fund's Beta in the papers, the Beta has likely changed again.

Statistics don't seem to help either when our hard-earned money is at risk. Calculating the average variance of a portfolio (i.e., the sum of the squares of the difference divided by the number of observations), the R-square factor or the standard deviation doesn't give much comfort to the average investor in a down market. Likewise, the modern portfolio theory of structuring a portfolio according to the efficient frontier concept, are not helpful when stocks, bonds, and or foreign funds take a sharp tumble. We can thus dispense with detailed discussions of how these accepted statistical entities relate to possible portfolio performance.

Another apparent flaw in the Beta theory is the universally accepted procedure to measure the Beta in terms of the performance of the overall stock market, for example, the S&P 500 index for U.S. investors. A fund manager's strategy to include in the fund's portfolio a certain percentage of so-called risk-free and less risky instruments, i.e., Treasury bills, cash, and bonds, to make stock market down-turns more palatable, runs contrary to our investment strategy. In the first place, there are no "risk-free" investments. Every investment type has an inherent risk, even 3-month Treasury bills or government guaranteed certificates of deposits (CDs).

Besides, we make our own decisions based on our own evaluation of the market. When market conditions tell us it's time to get off the wagon, we take decisive action by either completely eliminating or at least substantially reducing our portfolio's exposure in equity, bond, and/or foreign funds.

The difficulty that arises in this endeavor is the selection of the appropriate time span on which the risks and reward calculations are based. William F. Sharpe, American economist, has developed perhaps the most widely accepted method to determine risk, reward, and the expected return of a fund investment. However, the model developed by Sharpe seems to be specifically designed for equity portfolios whose Betas can be changed by diversification in form of using stocks with different Betas or by including "risk-free" Treasury bills in the portfolio. Again, this will not help in our strategy since we include, also at some time, foreign investments, bonds, and gold funds in our portfolio.

Notoriously, most fund managers are not very adept at recognizing market turns. They tend to be fully invested when the market turns down and underinvested when it turns up. They also do not have the freedom of the small investor to get in and out of the market all at once. Often, fund managers must guide their investment decisions not by market conditions but to meet heavy customer redemptions. Conversely, a fund might sometimes have unusually high cash inflows, forcing the manager to buy equities at a time when stock prices are on the overvalued side. In short, fund managers might be forced to sell or buy equities at the wrong time. These are risks that don't show up in conventional methods of risk analysis.

It is wrong to assume diversification into too many equity funds will reduce risk. At times they, as a group, could water down the overall return of the portfolio and, during severe market corrections, could all be hurt together at the same time. Not much portfolio risk control when one thinks about it.

5.5 Our Definition of Risk

From our point of view, we define risk as follows: risk is the likelihood of the portfolio returning less than the minimum required return. Returns less than the minimum required are losses. Returns higher than the minimum required are gains. To repeat: minimum return needed = (1 + inflation)/(1 – taxes).

Expressed in other words, risk is the potential loss in purchasing power of an investment due to inflation and/or taxes (also factoring in a possible drop in money market yields of 1%). The "minimum required return" is thus the break-even point between the profit and loss of a portfolio.

The following examples clarify this important statement

	Percent
Current inflation rate	4.5
Current tax rate	28.0
Minimum return required	7.6
Portfolio "A" return	5.4
Portfolio "B" Return	9.4

Hence from our viewpoint, Portfolio "A" showed a loss of –2.2% and Portfolio "B" a profit of 1.8% (subtracting the minimum required from the actual return). In essence, our risk definition places the main emphasis on preservation of capital. 100,000 today might have a purchasing power of only 30,000 in 25 years at an average inflation rate of 5%. For this reason, it is absolutely essential to structure a portfolio for a combined potential annual return that is consistently above the calculated minimum required return.

Here lies the difficulty in risk analysis. Most investors approaching retirement age tend to shift the majority of their investments into so-called "risk-free" secure income producing investments such as CDs, money market, or low risk bond funds whose returns do not meet our minimum required percentage. Any medium sized portfolio structured predominantly around such income producing investments is very vulnerable to erosion of its purchasing power and thus also susceptible to potential reductions in principal when a fixed amount is withdrawn monthly. Living off investment dividends and interests, without touching the principal, is possible only if the portfolio return is consistently above the minimum required year in and year out and the amount of withdrawals not higher than what is required to maintain the purchasing power of the portfolio over the first 20 years of retirement.

5.6 Life Expectancy

Many retirement plans designed by financial advisors are based on the average life expectancy of 71.4 years for men and 78.3 years for women. It would be wrong to calculate the monthly withdrawal amount based on these life expectancy levels. At age 65, a male investor will have a much greater chance to live another 15 to 20 years than to live only another 6.4 years until he reaches the suggested life expectancy. One should plan to need retirement withdrawals from the portfolio to age 90 to make sure the retirement nest egg will last and cover unexpected costs for 25 years of retirement living. This is not an unrealistic yardstick. It does not make sure there is something left over for the heirs, but does make sure we can cover any unexpected high health costs should we get seriously ill or need costly health care as we get older. We can never rely on the government to provide us this security.

5.7 Fund Type Risks

Now comes a very important part of our investment strategy. In essence, we must now determine the inherent, potential risk (loss) and potential reward (gain) for each of the main fund types. It is important to note that at no time will there be an attempt made here at predicting fund returns. All we want to do is to determine to what extent an investment is exposed to a potential loss or gain. The results obtained from this analysis are not a prediction but an indication of what could happen in a worst and best case scenario. Our risk analysis should not be confused with making projections.

No method, however powerful and sophisticated it might appear to be, will ever be able to precisely define the true risks and rewards of a portfolio. To state otherwise would be to deny reality. There are too many factors that cannot be measured such as factors that could govern price movements in securities, be that equities, bonds, or treasury issues. Nobody can foresee what kind of total returns these issues will deliver in the future.

Catastrophic and political events have in the past, and just as likely in the future, will adversely affect the markets. Wars, earthquakes, and financial debacles cannot be foreseen by anyone. No investor can be immune from such risk, for risk is part of investing.

5.8 Investment Advice Risk

One of the risks most often overlooked and very seldom considered is the investment advice risk. First, we look at these risks, then we discuss how our Fuzzy Logic approach defines risk. Most small investors, not familiar with the intricate aspects of the investment scene, rely on outside advice for their investment decisions. Many have worked all their lives and sacrificed a lot to acquire a sizable nest egg for their retirement. Most likely, they have led a prudent life style, adhered to a well-defined savings plan, and, sad to say, a great number of them didn't know how to invest their savings in a well thought out investment plan. For lack of expertise, they all too often entrust their life savings to a complete stranger because he has made promises of 20 or 30% annual returns. Or they entrust their money to a hot shot money manager who has made 65% profit for his clients the past year with the often false assumption he can repeat this performance for their nest egg as well. We say it again, beware of anybody who makes excessive promises or has only short term performance results to report. Know your investment advisor. Know his background and expertise. There are many deserving our business for they deal in an honest manner and make us aware of the risks; they even set our goals and objectives first before anything else. These are the type of managers who stay with the same company for a long time because they have built up a large base of satisfied customers.

But many boiler room operators are quick to take advantage of such naive investors to generate commissions for themselves. They usually don't stay for long in the same place. Some shady operators often churn the customer's account, get him into speculative high risk investments, or convince the investor to buy on margin. Relying on such questionable advice can be disastrous. Priority should be concentrated on providing the customer with superior service instead of placing the main emphasis on the amount of commission one can make from the inexperienced investor.

We have to add here also a positive note. Mutual fund companies in the U.S. show remarkable integrity and transparency in their dealings with their customers. Most of them also provide the superior service an investor expects.

We also caution against newspaper advertisements that promise high returns on investments. Also, don't accept a mutual fund brochure that is more than 6 months old regardless of how neat and glossy it is presented. Insist on the latest issue.

Also be aware of bank tellers playing the game of investment advisors. The future performance of the recommended investments very seldom will match the performance shown in the glossy and often outdated advertisement brochures they hand out to customers. Can you imagine a post office clerk selling mutual funds over the counter? A worker who perhaps doesn't even know how a mutual fund is managed and/or not having had the proper training and education to advise customers in investment matters. Absurd? Impossible? By no means, because this is exactly what the Swiss Postal Service, PTT, has recently started to do.

Imagine a customer going to the post office to buy stamps, to pay his taxes, telephone, and electricity bills (yes, they also provide bill paying services) and now doing mutual funds transactions all at the same time — in the post office of all places. Weird.

This is the type of risk an investor can easily eliminate by learning as much as he can about the investment markets. Frequent trips to the local public library can pay dividends many times over and can convert a novice into a prudent investor. Still, if outside advice is needed, one has to make sure to deal with a reputable advisor who understands one's investment objectives and goals. Most important, an advisor is needed who will structure the portfolio around the client's goals and not his own. An investor should never allow the advisor to talk him into an investment he doesn't understand. If the advisor is not willing or cannot explain the reasons for his recommendations and clearly define how this investment fits the investor's goals, the advice is not worth a dime. An investor should also keep in mind that an advisor or broker is a salesman first and foremost out to make a living for himself and a good one at that. Never should one make an investment decision based on a cold call from a broker or from a sales pitch by one of the boiler room operators over the phone. Hang up on him before he has a chance to dazzle you with figures and empty promises. And we repeat, never enter into speculative investments in the hope of making that once-in-a-lifetime big hit. Elimination of the investment advisor risk must be the first priority in risk analysis. Only after this has been taken care of can an investor concern himself in earnest with the actual task of risk management.

One must be particularly on guard against investing in things whose returns are based on a pyramid scheme where high profits are promised if the investor can lure other people into the plan. Such schemes are mostly offered over the phone and lately also over the Internet by fast talking con artists. The same pyramid schemes are often introduced by mail in form of introductory sample letters that should be sent to other people with the promise that each new investor joining the plan ("club") would generate a given reward for the ones that sent out letters on the lower tier of the pyramid. These are not investment plans or clubs; they are nothing else but a scheme to enrich the originators of these letters.

Then there are the real cold blooded operators that ask for a few hundred dollars or francs they need immediately so they could "free" a given large sum of money that is frozen in some place for some reason or another. Naturally, they promise to reward the donor with a certain percentage of the frozen money. This is one of the oldest scams known and it is amazing how unsuspecting people still fall for such ploys. These con artists have their eyes set mostly on elderly people because they know that this group is the most vulnerable. All these scams have one thing in common: the person who wants you to join or offers such a deal is always in a hurry for your money. They make one believe the promised rewards are forthcoming only if they receive your money right away. They will not allow time to think things over for obvious reasons. In such cases, the best answer is to let the con artist know that you don't have any money yourself but that your son who works for the police might be interested. Beware of many of these ploys. These people don't want you to invest — they want to get your money — all of it.

The U.S. stock market has delivered a historic average annual return of 10.7% since 1921. But, one cannot say that there is a likelihood an investor will obtain 10.7% annually if he buys a basket of stocks or a fund that mirrors the Dow Jones Industrials and holds it for the long run. No, it is not that easy. Using past long-term performances as yardsticks to project future returns is not devoid of pitfalls and dangers.

The same caution must be applied to the common belief that one has to take large risks in order to obtain large returns. Yes, mathematically, this has been proven many times over if one employs the conventional and commonly applied methods in calculating risk-to-reward ratios. As a counter argument we must add that this method fails to consider one important aspect of risk analysis, namely, the fact that the amount of risk an investment represents is directly related to the basic conditions prevailing at the time when an investment is being purchased.

Investment risks change if not weekly, certainly monthly. Risk is not a stagnant entity; it changes with every significant up or down move in the market.

For example, buying an equity fund is much more risky when the stock market is at a historic high and considerably less risky after the market has taken a sharp tumble. Equity funds react to market moves and stock markets might tumble in extreme bear market years, say 20 or 30%, or increase by the same amount in bull market years. Not so the individual equities. One particular stock might go all the way to zero and another might increase for decades every year by 20%. One can thus state that investing in the market (equity funds) is less risky than investing in individual stocks.

Likewise, a bond fund represents more risks when interest rates are in an up trend than when interest rates are trending down. Elaborating further, a foreign fund represents a high risk investment for a U.S. investor when the dollar gains in value against foreign currencies. Conversely, there is less risk in foreign investments when the dollar value falls.

We rely on the current economic and market conditions, determine if they are TOO HIGH-OK-TOO LOW, and make a buy, hold, or sell decision for mutual funds based on this evaluation. Risk control, for our purpose, is an extension of this Fuzzy Logic concept wherein we make sure our portfolio meets the criteria for the minimum required return. In other terms, we structure the portfolio such that its value will maintain, as a minimum requirement, its purchasing power irrespective of prevailing interest and/or inflation rates.

No investment should be purchased unless the associated risks are known. The majority of investors purchase funds with the focus on one single objective: namely, the potential profits in the future. To be a successful investor, one must be able to correlate the potential rewards with the prevailing risks any type of fund represents at the time the investment is intended to be made and also during the entire time the fund is held in the portfolio.

There is no time factor involved in our risk evaluation because, as mentioned, this is not an exercise in prediction. What follows is a purely empirical method based on the Fuzzy Logic control principle. We define herein what goes through our minds when we analyze a given situation and define in mathematical terms what could happen to our investments when certain assumptions could become reality. The key word here is "assumption," not predictions. Because they are assumptions only, there is no guaranty that these assumptions will materialize. However, from our own experience we have become convinced that this Fuzzy Logic approach to define risk is as good as or better than any of the other methods known today. The main concept in our evaluation is the recognition that any fund investment type follows a cyclic pattern where it could reverse its performance trend at any time to the down side (worst case) or to the up side (best case). The risks associated with any of the six main mutual fund types are shown in the next chapter. The values shown for the best and worst cases are not derived from historic stock or bond market index performances, but from hundreds of long-time mutual fund returns in each of the fund classes.

Our approach is not much different from the concept financial houses, investment advisors, and bankers are using but has different values assigned for what could happen to a fund in the best and the worst case situation. An important factor we consider is that mutual funds have less downside risks than an individual equity investment but also less upside potential than an individual high flying stock. Likewise, to high-income bond funds we assign much higher ranges between the best and worst case scenarios than for intermediate term government bond funds. The reader will immediately notice that the down side risk (worst case) for equity funds is much smaller than the historic market indexes such as the Dow Jones Industrials or the S&P 500 in a given year. Likewise, our down side risks for bond funds are less than some widely followed bond indexes. The reasons for these discrepancies are quite simple; first, most funds are structured around the objective of preservation of capital with each fund manager striving to limit the losses in a market downturn. In other words, a fund can only show loss levels investors are willing to accept. When the manager is not successful in this endeavor, investors will exit the fund in masses, and worse for him, his job might be on the line. The second and clearly most important reason is the Fuzzy Logic investment style we discuss in this book. By evaluating the current conditions of the markets by our TOO HIGH-OK-TOO LOW method, we should be able in most cases to exit given funds before markets turn sour in a major way. Last but not least, we are *not* short term traders (frequent switchers) or out to make a quick profit. Our relative long-term time horizon allows us to ignore daily and weekly market volatilities. These reasons combined will give our investment portfolio some added risk protection not found when we invest in individual securities alone.

For each of the main mutual fund groups we established constants for the best and worst case situation that will be used to determine the level of risk our current portfolio represents. This will be discussed in detail in the next chapter. We then compare the current portfolio risk with the personal risk level we have established for ourselves. This will enable us to decide if our portfolio is in need of an adjustment to bring its risk in line with our own personal risk level goals. The end result of this risk analysis will be a portfolio that according to our method of evaluation, will be structured in accordance to our personal needs. It will completely eliminate the need to have a financial advisor tell us what type of funds we should buy and how much should be invested in each of the major types of funds.

5.9 Investor's Personal Risk Level

Earlier we briefly mentioned the importance of determining the maximum risk level an investor could realistically assume. No risk definition, however sophisticated its mathematical and statistical background, will ever be useful for an investor unless consideration is being given to the investor's age, years until retirement, and his total investment portfolio size. No Beta, Alpha, and/or R-squared mutual fund risk indicator can be of much value unless these important entities are included in the risk analysis.

To our knowledge, no such indicator is used today by the majority of investors and investment advisors. The empirical formula developed by the author does take the above mentioned factors into account and is used in the herein described investment concept.

Axiom VI

The maximum risk level an investor should assume for the portfolio as a whole is

$$\text{Personal risk} = \frac{\text{Years to retirement}}{2} + \frac{\text{Portfolio size}}{50{,}000}$$

The lower the value, the lower the risk. Shorter times to retirement age justify lower risk values. Conversely, an investor in his early thirties can assume more risk with his portfolio than an investor who faces retirement in the next few years. Finally, the larger the portfolio size, the more risk an investor can assume.

Since there exists a direct relationship between portfolio returns and risks assumed, an investor structuring his portfolio around a high risk value also has the potential for above average returns in favorable market conditions. Naturally, the downside in declining markets can be just as dramatic because there is never such a thing as a free ride. The personal maximum risk level is an entity that will assume great significance when our method of asset allocation for mutual funds is discussed in the next chapter. In Table 5.1 a few examples of personal risk levels are shown using the above definitions.

Here are a few examples

Investor A: Portfolio size: 25,000; age: 53
Personal risk level = (12/2) + (25,000/50,000) = 7.

Investor B: Portfolio size: 500,000; age: 39
Personal risk level = (26/2) + (500,000/50,000) = 23.

Table 5.1 Investor's Personal Risk Levels

Years Until Retirement	*Portfolio Size (1000's)*						
	25	*50*	*75*	*100*	*150*	*250*	*500*
2	2	2	3	3	4	6	11
4	3	3	4	4	5	7	12
6	4	4	5	5	6	8	13
8	5	5	6	6	7	9	14
10	6	6	7	7	8	10	15
12	7	7	8	8	9	11	16
14	8	8	9	9	10	12	17
16	9	9	10	10	11	13	18
18	10	10	11	11	12	14	19
20	11	11	12	12	13	15	20
22	12	12	13	13	14	16	21
24	13	13	14	14	15	17	22
26	14	14	15	15	16	18	23
28	15	15	16	16	17	19	24
30	16	16	17	17	18	20	25
32	17	17	18	18	19	21	26
34	18	18	19	19	20	22	27
36	19	19	20	20	21	23	28

Risk Level = (Years to retirement/2) + (Portfolio size/50,000)

6 Asset Allocation

6.1 Modern Portfolio Analysis

How the individual fund groups are distributed in a portfolio is known as asset allocation. Our method of asset allocation differs from others. For example, American economist Harry Markowitz pioneered the concept of modern portfolio analysis. He proved that portfolios cannot be efficiently constructed with the sole purpose of maximizing future returns. He showed on mathematical models that consideration of risk was just as important as expected returns in portfolio selection. Moreover, he also demonstrated how the risk of a portfolio as a whole is different from the sum of the risks of individual investment types of which the portfolio is constructed. He introduced the concept of the *efficient frontier* in which a portfolio's potential return and risk are optimized to produce the most efficient trade-off between risk and return. However, this method has two serious deficiencies when applied to mutual fund investments. First, as with most other risk definitions, no consideration is given to the age of the investor as in our method. In other words, the Markowitz's method is applied identically to investors who have already retired and those who might still have 25 years to go until retirement. As discussed in the previous chapter, two investors of different ages could take risks quite different from each other. The other flaw is that Markowitz's method doesn't consider portfolio size in his definition of the *efficient frontier.* Both of the aforementioned factors however are of extreme importance to us.

We do not want to elaborate further on this widely accepted method of portfolio analysis because it requires mathematical skills, cumbersome calculations, and relies on past performance to anticipate future returns. In short, it is a method that doesn't agree with our own investment philosophy.

We do however fully agree with Markowitz's contention that risk is as important as rewards when analyzing portfolios. Nevertheless, we have come to the conclusion that there had to be an easier and more realistic way to determine an optimum portfolio asset allocation for specific individual needs. To this end, we have developed a method that will allow an investor to quickly determine if any portfolio meets the investor's maximum risk level depending on his age and size of assets. Most important, the thus developed method is based on current conditions, not on the distant past, nor on beliefs about the future. The new method introduced in this chapter meets all investment criterias discussed so far in this book and applies to individual mutual fund investors, or in other words, to portfolios exclusively made up from mutual funds. It does not apply to investors who buy individual stocks, bonds, or other investments, i.e., those investors who build their own portfolio from scratch. It is quite different from what is being taught in universities today.

As the reader has learned in the previous chapter, there is no time horizon in our method of risk analysis. The constants used are based on the assumption that a mutual fund has an equal chance to either reach the worst or best case scenario at any given time in the future. The question of when these investments will reach these assumed levels is irrevelant. It might happen next week, next month, or several years into the future — nobody knows.

Periodically one hears recommendations from experts specifying various mixes such as 50% equity, 40% bonds, and 10% cash. Where one investment advisor might recommend 80% equities, another at the same time might recommend 90% cash instead. Another might recommend to have two portfolios where one part will remain constant in its asset allocation structure over time. In the other part (known as the variable portfolio) assets are frequently changed in accordance to prevailing market conditions. This investment strategy doesn't appeal to us because such a twofold portfolio structure might contain assets that do not qualify as acceptable investments under given market conditions. In matters of asset allocation, divergent opinions always exist because bulls and bears want to have their say in this arena, too.

Let's view examples of recommended portfolio asset allocations by well-known investment advisors. This is to demonstrate how diversified the opinions generally are among their peers and, more important, how useless their recommendations can be with respect to an investor's personal risk tolerance. Just visualize a 75-year-old retired investor with a portfolio of less than 100,000 placing his assets the way any of these advisors have recommended. In hindsight, they were all dead wrong for this specific investor. Why? Because the examples given were all extracted from various financial publications in the same week, most noteworthy, in the spring of 1998, just months before the stock markets worldwide took a dive. These are clear examples of the pitfalls that exist today with the conventional risk analysis used by advisors based on current investment and "modern" portfolio management theories.

Examples of Recommended Portfolio Asset Allocations

Advisor	*A*	*B*	*C*	*D*	*E*
Aggressive Growth Fund	25	10	3	0	0
Equity-Income Fund	5	10	13	10	13
Intermediate/Long Term Bonds Funds	15	45	40	45	47
High Income Bonds Funds	5	10	4	0	0
Foreign Funds	25	15	30	15	19
Money Market Funds	25	10	10	30	21
Worst	–11	–9.4	–13.5	–6.8	–8.9
Best	23.9	21.0	25.3	18.0	20.2
Portfolio Risk Level	46	40	52	32	38

Worst, Best, and Portfolio Risk Level data in the table above were calculated by our method discussed in detail below.

We notice immediately the relative high risk levels of all recommended portfolios. According to our own concept of portfolio management, the above recommendations are suitable only for investors with very large portfolios and/or with many years to go until retirement. We can only guess what kind of explanations these advisors will have when this 75-year-old investor, with a portfolio of $100,000, followed their advice and the stock and bond markets tumbled. (Note: according to our concept, this investor's risk level was **2**). Is it going to be: "Don't worry, the markets will come back again," or is it: "Don't sell, it's only a paper loss"? One answer this investor will never hear is this: "We're sorry — we goofed. We should have taken your personal financial situation more under consideration and thus will credit your account for the amount of loss you suffered because of our mistake."

One advantage we have in structuring the portfolio allocations ourselves is that we do not have to listen to bankers, brokers, and investment advisors. We don't want to join the herd when this group of so-called experts starts to get greedy, overly confident, and reckless in handling other people's money. They see the competitors have done better than themselves and thus, to stay competitive, they adapt the same investment techniques as their peers, irrespective of risks involved. When a market guru buys a given stock, they follow suit and, in doing so, drive the price of the stock up. Then they all sell at the same time, leaving the investor who believed in their better stock picking "expertise" in the dust and holding the short end of the stick. It takes courage to form one's own opinion and to stick to one's convictions. In the past few years many of these hot shot managers might have laughed at our conservative investment style, but the 1990, 1997, and 1998 corrections have shown our concept to be viable.

Naturally, no investment advisor, banker, broker, mutual fund company, or even this author can ever guarantee a no-loss performance at any one time, but there clearly has to be more prudence shown toward an investor's own needs and less emphasis placed on commissions and fees generated. A two-way partnership for profits on both sides is the only course that will have a chance to endure.

For each of the main mutual fund groups, we established constants for the Best and Worst case situation as shown in Table 6.1.

Table 6.1 Mutual Funds Risk Levels

	Worst Case	*Best Case*
Money Market funds	–1	+1
Aggressive Growth funds	–10	24
Equity-Income funds	–6	20
Intermediate and long term bond funds	–4	15
High Income bond funds	–10	19
Foreign Funds	–35	48

It is important to note that we let these numbers remain constant irrespective of market conditions. Important: these are not to be viewed as predictions for future returns. Their sole purpose is to assign a risk surrogate to each of the different fund types.

6.2 Portfolio Risk Level

For a quick overview, we have developed computer-driven spreadsheets as shown in Tables 6.2 to 6.4. These can be used by the reader as aids to view different asset allocation models in the low, medium, and high risk classes without the need for him to do the elaborate calculations himself.

Table 6.2 Asset Allocation Models (Low Risk)

Current Money Market Yield: 4.9 Maximum Risk Level: 9

	Money Market	*Aggr. Growth*	*Equity-Income*	*Interm. Bond*	*High Inc. Bond*	*Foreign Fund*	*Portfolio Worst*	*Portfolio Best*	*Portfolio Risk*
** →	90			10			3.1	6.8	1
** →	90		5	5			3.0	7.1	1
** →	90		10				2.9	7.3	1
** →	90	5		5			2.8	7.3	2
** →	85			15			2.7	7.3	2
** →	90	5	5				2.7	7.5	2
** →	90	5	5				2.7	7.5	2
** →	85		5	10			2.6	7.5	2
** →	90	10					2.5	7.7	3
** →	85		10	5			2.5	7.8	3
** →	85	5	5	5			2.3	8.0	3
** →	80		5	15			2.2	8.0	4
** →	85		10		5		2.2	8.0	4
** →	85	5	10				2.2	8.2	4
** →	85	10		5			2.1	8.2	4
** →	80	5		15			2.0	8.2	4
** →	80		5	10	5		1.9	8.2	4
** →	85	10	5				2.0	8.4	4
** →	80		15	5			2.0	8.5	4
** →	85	10			5		1.8	8.4	5
** →	80	10		10			1.7	8.6	5
** →	75		10	15			1.7	8.7	5
** →	90			5		5	1.6	8.5	5
** →	75	5		20			1.6	8.6	5
** →	80	5	5	5	5		1.6	8.6	5
** →	75		10	10	5		1.4	8.9	6
** →	80	10	10				1.5	9.1	6
** →	90				5	5	1.3	8.7	6
** →	75	5		15	5		1.3	8.8	6
** →	80	15		5			1.4	9.1	6
** →	85			10		5	1.2	8.9	7
** →	75	10	5	10			1.2	9.3	7
** →	70		15	15			1.2	9.4	7
** →	75	10	10	5			1.1	9.6	7
** →	70		20	10			1.1	9.6	7
** →	70	5	10	15			1.0	9.6	8

Table 6.2 (continued) Asset Allocation Models (Low Risk)

		Money Market	*Aggr. Growth*	*Equity-Income*	*Interm. Bond*	*High Inc. Bond*	*Foreign Fund*	*Portfolio*		
								Worst	*Best*	*Risk*
**	→	65		10	25			0.9	9.6	8
**	→	70	10	5	15			0.8	9.8	8
**	→	80		5	10		5	0.7	9.6	8
**	→	75	10	5		10		0.6	9.7	8
**	→	60		5	35			0.6	9.8	9
**	→	75	20		5			0.7	10.0	9
**	→	80		10	5		5	0.6	9.9	9
**		75			20		5	0.4	9.8	9
**		70	15		10	5		0.3	10.2	10
**		75	25					0.4	10.4	10
**		65	5	20	10			0.4	10.5	10
**		80		5		10	5	0.1	10.0	10
**		90					10	0.0	10.1	10
**		55		10	35			0.1	10.5	10
**		70	20	5	5			0.2	10.7	10
**		60		25	15			0.2	10.8	10
**		70		5	20		5	–0.1	10.5	11
**		70	25		5			0.0	10.9	11
**		60	10	10	20			–0.1	10.9	11
**		55		20	25			–0.1	11.0	11
**		55	5	10	30			–0.2	10.9	11
**		75	5	15			5	–0.2	11.0	11

On the tables the left portion lists various portfolio mixes, each number representing the percentage of a given fund classification (Money Market, Aggressive Growth, Equity-Income, Intermediate Bond, High Income Bond, and Foreign Fund) contained in the portfolio. All mixes are arbitrarily chosen alternatives of asset allocation. The reader might want to expand this table to more and different alternatives or he might want to limit a given fund group to a given percentage of the portfolio as a whole. For example, an investor might have a policy to limit foreign fund holdings to less than 5% and equity funds to not more than 40%.

The ** signs to the left of the table are pointers for the portfolio mixes whose Risk Levels are below the ones indicated at the top of the table. In our computer program, these stars are inserted automatically by an "IF" command.

Table 6.3 Asset Allocation Models (Medium Risk)

Current Money Market Yield: 4.9 Maximum Risk Level: 14

	Money Market	*Aggr. Growth*	*Equity-Income*	*Interm. Bond*	*High Inc. Bond*	*Foreign Fund*	*Portfolio*		
							Worst	*Best*	*Risk*
** →	60	10	15	15			–0.2	11.2	12
** →	55		25	20			–0.2	11.2	12
** →	70		15	10		5	–0.3	11.0	12
** →	65	25		10			–0.4	11.3	12
** →	75	10	5		5	5	–0.6	11.2	12
** →	60	10	20	5	5		–0.6	11.6	13
** →	65	30		5			–0.7	11.8	13
** →	55	15	10	20			–0.8	11.8	13
** →	70	15		10		5	–0.9	11.6	13
** →	55	20		25			–0.9	11.8	14
** →	60	20	5	10	5		–0.9	11.8	14
** →	50	15		35			–1.0	11.8	14
**	60	25	10	5			–1.0	12.3	14
**	60	30		10			–1.1	12.2	14
**	55		10	30		5	–1.4	12.1	15
**	50	20	5	25			–1.4	12.5	15
**	45	10	15	30			–1.3	12.6	15
**	55	25		15	5		–1.5	12.4	15
**	60	35		5			–1.4	12.7	15
**	40		20	35	5		–1.5	12.6	16
**	50	15	15	15	5		–1.6	12.8	16
**	40	5	20	35			–1.5	12.8	16
**	45	10	5	30	10		–1.7	12.5	16
**	50		10	35		5	–1.8	12.6	16
**	60	15		15	5	5	–2.0	12.7	17
**	75	15				10	–2.1	12.8	17
**	55		15	15	10	5	–2.1	12.8	17
**	40	5	35	20			–1.8	13.6	17
**	30		20	50			–2.0	13.3	17
**	70	10	5	5		10	–2.3	13.1	18
**	55	40		5			–2.1	13.6	18
**	40	20		35	5		–2.3	13.4	18
**	45	5	5	40		5	–2.4	13.3	18
**	35	10	10	40	5		–2.3	13.4	18
**	35	10	15	35	5		–2.4	13.7	19
**	50	30	10	5	5		–2.4	13.9	19

Table 6.3 (Continued) Asset Allocation Models (Medium Risk)

	Money Market	*Aggr. Growth*	*Equity-Income*	*Interm. Bond*	*High Inc. Bond*	*Foreign Fund*	*Portfolio*		
							Worst	*Best*	*Risk*
**	45	30	10	15			–2.4	14.1	19
**	60	30		5		5	–2.6	13.9	19
**	40	15	25	15	5		–2.5	14.2	19
**	30	10	10	45	5		–2.7	13.9	19
**	35	15	25	25			–2.6	14.4	20
**	35	25	5	35			–2.8	14.3	20
**	55	20	20			5	–2.8	14.4	20
**	40	25	5	20	10		–3.0	14.3	20
**	55	30		10		5	–3.0	14.3	20
**	40		20	30	5	5	–3.1	14.2	20
**	40	30	15	15			–2.9	14.8	21
**	60	5	25			10	–3.2	14.5	21
**	30	10	30	25	5		–3.1	14.9	21
**	25		50	25			–3.0	15.2	21
**	60	10	10	5	5	10	–3.5	14.4	21
**	35	5	15	40		5	–3.4	14.7	21
**	50		20	20		10	–3.6	14.8	22
**	45	5	25	5	15	5	–3.7	14.9	22
**	25	10	20	35	10		–3.6	15.0	22
**	35	5	25	30		5	–3.6	15.2	22
**	35	10	15	35		5	–3.7	15.1	22

Tables 6.2 to 6.4 show the minimum return needed and the investor's maximum Risk Level, both determined as discussed in the previous chapter. (Naturally, each investor would enter his own risk level and current money market yield here.)

In our examples (Tables 6.2 to 6.4) we have reduced the multiple variation possibilities to be able to show the most important ones on a single page. Our actual asset allocation model contains at least 50 more variations, the number of choices being strictly a matter of preference.

Numbers indicated in the Worst and Best columns on the right refer to the established constants in Table 6.1 applied to the portfolio as a whole and calculated as follows:

Table 6.4 Asset Allocation Models (High Risk)

Current Money Market Yield: 4.9 Maximum Risk Level: 25

	Money Market	Aggr. Growth	Equity-Income	Interm. Bond	High Inc. Bond	Foreign Fund	Portfolio		
							Worst	Best	Risk
** →	45	25	10	15		5	–3.7	15.3	23
** →	55	10	20	5		10	–3.8	15.2	23
** →	45		20	25		10	–3.9	15.2	23
** →	25	25	15	35			–3.8	15.7	23
** →	30	30		30	10		–4.0	15.4	23
** →	40	10	40	5		5	–3.8	15.9	23
** →	45	25	10	10	5	5	–4.0	15.5	23
** →	40	50		10			–3.8	15.9	24
** →	40	20		25	10	5	–4.2	15.2	24
** →	50	40		5		5	–4.0	15.7	24
** →	45	25	15	5	5	5	–4.1	15.8	24
** →	40	30		25		5	–4.2	15.7	24
** →	20	30		50			–4.2	15.9	24
** →	30	10	40		20		–4.2	16.0	24
**	30	10	30	25		5	–4.4	16.3	25
**	35	50	5	10			–4.3	16.6	25
**	50	25		15		10	–4.7	16.0	25
**	25	20	40	10	5		–4.3	16.7	25
**	45	40	5	5		5	–4.5	16.4	25
**	20	30	15	35			–4.5	16.6	26
**	25	20		50		5	–4.8	16.2	26
**	25	40	10	20	5		–4.9	17.0	27
**	35	60		5			–4.8	17.2	27
**	30	50	10	10			–4.8	17.3	27
**	25	40	20	15			–4.8	17.3	27
**	45	20		15	10	10	–5.3	16.4	27
**	45	20		15	10	10	–5.3	16.4	27
**	40	10	30	10		10	–5.1	17.1	27
**	35	40		20		5	–5.2	17.1	27
**	30	45	15		10		–5.2	17.5	28
**	35	10	25	20		10	–5.4	17.3	28
**	25	50	10	15			–5.2	17.7	28
**	35	30	10	10	10	5	–5.4	17.1	28
**	25		20	45		10	–5.5	17.0	28
**	35	10	15	20	10	10	–5.8	17.2	29
**	30	10	20	30		10	–5.7	17.5	29

Table 6.4 (Continued) Asset Allocation Models (High Risk)

	Money Market	*Aggr. Growth*	*Equity-Income*	*Interm. Bond*	*High Inc. Bond*	*Foreign Fund*	*Portfolio*		
							Worst	*Best*	*Risk*
**	25	10	10	45		10	–5.9	17.4	29
**	25	60		15			–5.6	18.1	29
**	35	10	30	10	5	10	–5.8	17.7	29
**	15	10	30	35	5	5	–5.9	17.9	30
**	20	10		60		10	–6.1	17.4	30
**	25		40	25		10	–5.9	18.0	30
**	30	10	10	25	15	10	–6.4	17.6	30
**	35	20	5	20	10	10	–6.2	17.6	30
**	35	20	20	10	5	10	–6.2	18.1	31
**	20			50	20	10	–6.7	17.3	31
**	25	40	10	20		5	–6.2	18.5	31
**	20	10	35	10	20	5	–6.5	18.3	31
**	30	20	20	20		10	–6.3	18.4	31
**	40	40	10			10	–6.5	18.8	32
**	25	50		20		5	–6.6	18.9	32
**	35	20	20		15	10	–6.8	18.5	32
**	20	25	30	10	10	5	–6.7	19.0	32
**	25	20	10	30	5	10	–6.8	18.5	32
**	20	30	20	10	15	5	–7.1	19.1	33
**	15	10	20	45		10	–6.9	18.8	33
**	20	20	15	35		10	–7.0	19.0	33
**	30	30	5	15	10	10	–7.2	18.9	33
**	30	10	20	20	5	15	–7.6	19.3	34
**	20	40	25	5	5	5	–7.2	19.9	34
**	15	45	5	30		5	–7.2	19.6	34
**	15	50	10	20		5	–7.6	20.3	35
**	20	20	40	10		10	–7.5	20.3	35

$$\text{Worst} = a*(X-1)+b*(-10)+c*(-6)+d*(-4)+e*(-10)+f*(-35) \quad 6.2.1$$

$$\text{Best} = a*(X+1)+b*(24)+c*(20)+d*(15)+e*(19)+f*(48) \quad 6.2.2$$

where X = Current Yield on Money Market Funds (%)
a = % of Portfolio Assets in money market funds
b = % of Portfolio Assets in aggressive growth funds
c = % of Portfolio Assets in equity-income funds
d = % of Portfolio Assets in interm./long term bond funds
e = % of Portfolio Assets in high income bond funds
f = % of Portfolio Assets in foreign funds

The portfolio risk level, as its name implies, is an indicator of the portfolio's risk as a whole. Again, this is an index that is purely empirical in nature, calculated as follows:

$$\text{Portfolio Risk Level} = (\text{Best} - \text{Worst}) - \text{Worst} \qquad 6.2.3$$

As mentioned earlier, each investor has his own risk tolerance. Clearly, an investor in his late twenties can assume more risk than a retiree who must live off the income of his portfolio. The level of risk an investor could assume is therefore related to the length of time he has until he plans to withdraw money from the portfolio. The size of the portfolio, as we have mentioned, is also a governing factor. Someone with a portfolio of say 1.5 million, even at age 70, clearly can stand more risk than an investor with "only" 150,000 and two years to go until retirement.

One of the real positive aspects of limiting an investor's risk to a given level is that it will act as a constraint, preventing a mutual fund investor from using a speculative and too aggressive investment style. Conversely, it also will prevent an investor from having a too convervative portfolio structure.

The selection process for the "optimum portfolio" mix is not difficult. First and foremost, an investor will enter into the table the mix he currently holds in his portfolio. He then compares his own portfolio Worst, Best, and Risk Level with the ones that are marked **.

Within this group of marked mixes might be one that

1. has possibly a better Worst
2. higher Best and/or
3. a lower Risk Level than his own.

There might also be special times, unique to an investor's situation, that could compel him to use a less risky portfolio than what his personal risk level would call for. For example, he might just have an aversion to investing in either foreign or aggressive equity funds. Or he feels so uncomfortable with his portfolio that he might lose sleep over it and worry day and night how his funds are doing. There is no reason to sacrifice one's own well-being for such a mundane thing as investing. Life is too short to become a slave to money matters. By all means, an investor should then select a less risky portfolio if that would make him enjoy life more.

One should not get the notion to scramble the portfolio around any time one's own mix is not quite as close to the ones indicated by the ** markers. Having a so-called out-of-line portfolio mix doesn't mean it has to be changed right away. There is no great urgency here. In such a case, the investor only knows that there are better alternatives available. And with this information as a starting point, he must now do the final steps. Decisions for buying and selling portfolio assets will become for him a question of proper timing. For this he will use the Fuzzy Logic method discussed in detail in the next chapter.

USING FUZZY LOGIC

7 Fuzzy Logic: The TOO HIGH-OK-TOO LOW Concept

For ages, investors have attempted to find the magic indicators that could forecast future price movements of investment securities. To date, no one has achieved this goal and we can safely assume no one ever will. Many advisors claim to have the golden touch to pick the future winners in the investment game. It's an illusion; they're deceiving themselves. Investors spend a lot of money for this kind of advice and more often than not, the end results are not much better than if they had made the investment selections for themselves. These advisory services are very adept at showing in elaborate detail how individual investment securities or mutual funds have performed in the past, but when it comes to predictions on their part, they must be taken with a grain of salt. It doesn't mean these services are useless, by no means; this kind of information is quite important to investors, but they are not the end to all means.

There are perhaps as many investment styles as there are investors. Likewise, the many types of analyses and indicators available today to investors could fill volumes. Modern telecommunication technology has made this possible. Anybody can now get just about any information needed from around the world in a split second. We can even be on a cruise ship, a thousand kilometers away from our home base and find out at what price our investment has traded on the stock exchanges fifteen minutes earlier. Hardly any month goes by without someone claiming to have found a new, more reliable indicator to foretell market moves. Computers have made it possible to process tons of data in matters of seconds and, as a result, the investment scene has become saturated with a multitude of fad indicators that tend to quickly disappear into obscurity after the initial euphoria has worn off. First let's look at some of the indicators used to analyze the market, then discuss the indicators we use for the Fuzzy Logic method.

7.1 Technical Analysis of Markets

The Technicians believe future price directions can be predicted based on past data. They adhere to the theory that price movements follow certain patterns that tend to repeat themselves over time. They use charts, measure angles, draw trend lines, and calculate moving averages, etc., etc. Problems arise when the trend line is broken or when the support or resistance levels have been penetrated. Then it's back to the drawing board, a new line is drawn and, with much fanfare, a "new trend" is proclaimed. Unless, a week later, the new trend reverses itself and becomes the old trend again. And we can be sure that the Technician will be able to present a perfect explanation for this abberation by drawing another line or perhaps a triangle on a piece of paper. Triangular relationships are great for compositions in paintings. Michelangelo used it and so did Rembrandt, but since when do harmonious relationships exist in the financial markets?

Then, there are the cycle worshippers who, instead of straight lines, draw waves — small ones and banzai type whoppers. They believe cyclical price movements repeat themselves over and over again into infinity. There is the Kondratieff cycle (50-54 years), the business cycle (4-6 years), the election cycle (4 years), etc., etc.

7.2 Fundamental Analysis of Markets

Fundamentalists concentrate on finding undervalued investments that have the potential for significant price appreciation in the future. They deal with indicators such as P/E ratios, earnings growth, price-to-book ratios, cash flow, debt-to-equity ratios etc., etc. Most mutual fund managers fall into this group. One only has to read a few fund prospectuses to learn that most funds have the objectives of:

> "Capital appreciation by investing mainly in common stocks of companies that have above-average growth potential,"

"Invest in stocks that may be undervalued, overlooked, or out-of-favor,"
"Invest in stocks of companies with valuable assets or in companies believed to be undervalued based on company assets, earnings, or growth potential."

One immediately notices that the emphasis is on the words "believed," "may be," or "potential." They have something else in common: they are mostly right in a bull market (it's easy to be a hero when the market goes up) and mostly dead wrong when the market tumbles.

7.3 Contrarian Investors

Contrarians profess to go against the main stream. In other words, when everybody buys, they sell, and vice-versa. They adhere to the theory that when investor sentiment is predominantly bullish, the market is ripe for a correction. Conversely, when everybody is bearish, it's time for the contrarians to buy. Most contrarians are both Technicians and Fundamentalists except that their wires are crossed. Contrarians pay special attention to market psychology and use sentiment indicators for their investment decisions.

7.4 Unusual Indicators

Finally, there are quite a few strange and exotic indicators. There are people who seem to have found the magic connection of all kinds of weird events with the stock market. For example, the Hemline Indicator is supposed to foretell the future market movements because as they say, when the hemlines in ladies fashions go up, the market will go down. And naturally, when hemlines go down, the market should go up. Or is it the other way around? Luckily Scotland and Tahiti don't have a major stock exchange; otherwise, those markets would never change.

The Superbowl Indicator is another one of these exotics. AFC wins, market will go down that year. NFC wins, a good market year will follow. For our readers overseas not familiar with these terms: this indicator relates to the championship game in American football.

What will they dream up next? A soccer indicator for the DAX (German Stock Exchange) or a cricket indicator for the FT-100 (England)? Then there are also some investors who let astrology guide them in their investment decisions. For them, the position of Mars, Jupiter, Saturn, etc. becomes an important factor for future market movements. Lunar cycles are thrown in for good measure. An astrology buff might tell us in all sincerity that a bull market, born in the month of June (Gemini) is in for a rough ride and a lot of volatility — because everybody is supposed to know that Geminis have large mood swings, are quick to change their mind, and cannot sit still for prolonged periods of time.

It is not our intention to make fun of all these indicators and methods of investment analysis. Neither do we want to give the impression that they are all useless. Some are, but many are not. The fact is that each and every one of the hundreds of indicators used in the financial community today, at some time or another, was correct in predicting future market movements. Otherwise some of these well-known indicators wouldn't have gathered such a large following. They all represent different opinions, generate different investment decisions among millions of investors, and, thus, make the market the way it is today — unpredictable and not following any set rules.

An investor is however well-advised to keep in mind that:

1. There is no single indicator that can be used to accurately forecast market movements six or twelve months into the future.
2. Relying on one single indicator for investment decisions can produce results similar to rolling the dice in a casino.
3. Sometimes history repeats itself, but sometimes it doesn't. Hence, any indicator that is based on past events and used to predict the future has about a 50/50 chance to be correct.
4. Any financial advisor who forecasts a specified price level at a specified future date is likely to reap more financial gains for himself than for the investor who follows his predictions.
5. Using a vast number and great assortment of indicators doesn't improve the reliability of a forecast. Following 30, 100, or more indicators tends to cloud the forecast and cause conflicting signals.

Investors and advisors alike use indicators to forecast the future. Some become so convinced of a particular indicator's reliability that they base all their decisions on that indicator alone without any other considerations. There are also others who believe vehemently in a specific indicator but dismiss any unusual and unconventional signal as an abberation.

Here are a few actual examples extracted from the financial news media to demonstrate the mentality that often prevails among "market experts." Quote:

> "My projections are that stocks (S&P 500) will appreciate 30% in the next 12 months while corporate bonds should decline about 11%. Fixed income instruments carry much greater risk than stocks in the current enviroment."
>
> "This is one of the strongest relative strength performers in the Dow 30. Although temporarily overextended, the major uptrend is still clearly intact. There's solid short-term support in the low 90's."
>
> "The market outlook has improved significantly since our last issue. Our model now projects that the S&P 500 index will advance 18% during the next 12 months. The recent correction improved the condition of several key indicators and helped boost our forecast."
>
> "Right now, mutual funds have an average of 9.3% in cash, and this indicator could go as low as 5% before the market is vulnerable. Insider buying is impressive, too; currently this signal is bullish."

The reader might have guessed right. Yes, these words of wisdom were written by some well-known market "experts" just a few weeks before the October 1987 crash. Need we say more?

Everybody is allowed his own opinion, but when an advisor issues a claim for an impending imminent boom or bust, investors must make sure not to take these "people in the know" for real. Many of these predictions might have a certain entertainment value; they can make us laugh sometimes. These are no laughing matters when we take them seriously and our portfolio will be hurt because we followed this advice. Our strategy demands a different kind of thought process to be used when we look at any of our indicators; namely, in our concept, indicators are tools to determine the current conditions only and are not to be used for forecasting the future.

One of the advantages of investing in mutual funds is that we don't have to be concerned about indicators that relate to individual stocks, bonds, or other securities. We can let the fund manager worry about these things; after all, we pay him a management fee to pick the winners. Odd Lot Sales, Put/Call Ratios, Intrinsic Stock Valuation, Momentum, or other indicators are his domain and do not concern us.

We use our own indicators to tell us nothing else than the *current* state of the economy, the government's fiscal policy, and the *current* state of the equity, bond, and foreign investment markets. The indicators tell us if present conditions are TOO HIGH-OK-TOO LOW. We use this TOO HIGH-OK-TOO LOW terminology throughout this book to remain consistent, but it could easily be replaced by the term triads "increasing-stable-decreasing" or "up-same-down."

7.5 Factors Influencing a Nation's Economy

To better understand the indicators for the economy discussed in 7.6.1, 7.6.2, and 7.6.3, we must become familiar with the cyclic behavior of the economy. Here, we are not talking about a cycle within a given time frame, but the fact that each economy passes through periods of up and down moves. This cyclic behavior of the economy is an ironclad occurance everywhere. There has never been an economy that has moved upward unrestrained forever without a setback. A nation's economy must go through periodic downturns to eliminate the excesses associated with a boom. Likewise, there will always be a recovery somewhere in the future that will lift an economy out of a recession. Changes will occur with guaranteed regularity, but nobody knows when. It could take weeks, months, or years for a change. It does so in the U.S., in Switzerland, Japan, or for that matter any other place on earth. It happens not only to a country, but also to states, cities, and even common households.

Money is the driving force, the engine, for any economy. Sometimes, there is too much money around; sometimes not enough. Modern society wants prosperity and will not accept long periods of hardship and belt tightening. There is a mechanism in place that allows for "greasing" the economy, wherein governments and individuals can get their hands on things they want now but for which they don't have the money to pay. This mechanism is known as credit. Private individuals and corporations know that sooner or later they have to pay back that debt and in doing so, know also such debt repayment will likely hurt their future spending plans. Excercising fiscal responsibility means making tough budget decisions because a debt has to be paid back in one way or another, no ifs and no buts. And so it should be.

There is nothing wrong with someone taking a loan in difficult times to put food on the table, feed the animals, or to have a shelter for one's children. There is also nothing wrong when a company issues debt securities to allow it to pay wages and its suppliers during hard times. However, they all know the golden rule of using credit that states: a debtor must reduce his debt load when the economic and financial conditions have improved again.

An even better solution would be to have no need for credit at all by setting aside money in good times for any possible emergency in the future. Unfortunately, all too many politicians adhere to the belief that governments are exempt from having to exercise fiscal restraints. Having the use of a printing press to create money out of thin air and also the ultimate power to regulate who pays and who receives money can often lead to excesses that directly will have an influence on a nation's economy and fiscal condition.

Only governments can control how much money there will be in circulation and what group will benefit most from this money distribution. Proper remedial action to improve the government's financial condition either comes too late or at the wrong time because political priorities might be focused on something else. This can lead to chaos in the economy and is a prime reason why the economy undergoes cyclic patterns. Not because the population has changed its spending habits, not because bankers have raised interest rates, and also not because industry had to raise prices for goods produced. No, all these are only secondary reactions to previous failed efforts by the government to manage the economy.

Each change in economic conditions has different causes and presents different situations to be mended. There is no remedy to cure all economic woes. The basic John Maynard Keynes (British economist) solution to throw more money on the street works sometimes, but sometimes it will not. Providing foreign countries or domestic companies with loans to bail them out will not help if the loan recipients won't change the ways that got them into the financial mess in the first place. This is why we assess the current economic conditions only for our investment decisions and do not rely on economists and government officials to tell us what actions and measures they will take to fix the economy.

7.6 Indicators to Assess the Current Economy

To assess the current state of the economy, one has to make use of indicators that are revelant and reliable. Taking a glance into the financial section of a newspaper, we might come across a whole series of indicators directly related to a nation's economic state. There are numbers for unemployment, factory inventory, sales, goods produced, etc., etc. In fact, there are so many indicators one could easily fail to see the real economic trend. Now let's look at the indicators we stated in the Preface to assess the current economy, the first step in our Fuzzy Logic method of analysis.

7.6.1 The Gross National Product Indicator

For our purpose, the first indicator we use is the change in Gross National Product (GNP) figure published monthly or quarterly by the government. The GNP is the total of all goods and services produced and reported as a percent change from the last report. In short, if the GNP figure is lower than before, the nation's economy has contracted; if higher, the economy has expanded. A recession is generally defined as a period in which the GNP has been negative for two consecutive quarters. In our Fuzzy Logic concept we consider any change of more than +0.4% as high, lower than –0.4% as low, and any change between this range as OK. Thus, in our terminology, the GNP can be either High, OK, or Low.

7.6.2 The Inflation Rate Indicator

The second indicator to assess the economy is the current inflation rate. Inflation has been defined as that condition in which too much money chases too few goods. An economy can become overheated or the central banks can flood the market with too much newly printed money such that paper money is available, but not enough goods to meet the demand. Result, prices will go up, inflation rate increases. With goods and services becoming more expensive, workers will now demand higher wages, causing prices to increase further. End result: even higher inflation.

And this spiral is continuous until the central banks decide to tighten, i.e., to restrict, the money supply to get inflation under control. It would take a wizard to time and manage the money supply such that the economy would not overheat nor slip into a recession. This is the job of the Federal Reserve Bank of the U.S., known in other countries as the central bank. The head of this bank is one of the most powerful men in government to guide the economy. A country could possibly function quite well with a mediocre President, Bundeskanzler, or Prime Minister, but it cannot function with a mediocre Chairman of the Federal Reserve, Chancellor of the Exchequer, or President der Bundesbank. Luckily, most central banks are an independent arm of the government whose decisions in money supply matters cannot be dictated by elected politicians.

The appointed members of this board are mostly former bankers and thus have the basic understanding of macro economics and so it should be.

A trend change in inflation rate of more than +0.4% is considered High, a reduction of more than –0.4 as Low, and the range in between as OK.

7.6.3 The Federal Fund Rate Indicator

The earliest indication that tells us of government interference in the money supply is when the governing central bank changes its key interest rate. In the U.S. this would be the case when the Federal Reserve Bank changes the Federal Fund Rate. In Germany it would be the Bundesbank changing the Lombard rate. When this rate falls, it is a sign of the central bank easing; when it rises, a sign of tightening in the money supply. Easing means more money is being made available or, in other words, the central banks are trying to give the economy a boost.

Confirmed long term trend changes in the Federal Fund Rate (Fed Fund Rate) will produce a like reaction in all the other interest rates that affect the corporate, banking, and private sector. In short, by changing the Fed Fund Rate, the Central Banker's aim is to manipulate the interest rates. Loosening the printing presses (placing more money into circulation) will lower interest rates. Tightening (reducing the money supply) will raise the interest rates.

The Fed Fund Rate then is our last indicator to assess the current state of the economy. Like with inflation, we consider a change of more than +0.4% as High, a lowering of more than –0.4% as Low, and the range between as OK.

7.7 The Fuzzy Logic Concept

Let us explain using a simple example how we interpret our thoughts in this Fuzzy Logic process. Presume we are driving our car at sixty miles an hour through a small community at night. Now, we know that our present speed is too fast for this kind of condition. Too fast for what? There might be a radar control, a group of pedestrians, or a tree across the road just around the corner ahead. Another driver might keep on barreling down the road at this speed under the same condition because he thinks the conditions are OK; after all, there couldn't possibly be a policeman around and no one else on the road this late at night. Our thoughts however are different. Because we don't know what lies ahead, considering the possibilities, our current speed for present conditions is too fast, so we slow down. This is to guard ourselves from any potential pitfall that might arise in the future. The reader will not have any difficulty recognizing the similarities in thoughts that might go through a driver's and an investor's mind.

7.8 Fuzzy Logic to Assess the Current Economy

We thus employ three universal indicators to give us a read of the current state of the economy. They are

Gross national product (GNP),
Inflation rate, and
Fed Fund rate.

These three entities will never be in equilibrium for long periods of time. Fluctuations will occur, regardless of the central bank's attempt to hold them as steady as possible. These variables move in undetermined cycles over time. Since each of these indicators can be High, OK, or Low, we will have thus $3 \times 3 \times 3 = 27$ **basic conditions** for the economy that can prevail at any time. These 27 basic conditions are represented graphically in Figure 7.1 (Fuzzy Logic Path for the Economy).

Figure 7.1 illustrates the economic cycles that tend to repeat themselves in periodic intervals. Let's use an example to explain the concept of a Fuzzy Logic Trail. We start with a recession identified as Condition 1 where GNP = low, Inflation = low, Federal Fund Rate = high. Assume now that the Federal Reserve Bank decides to boost the economy by lowering the Fed Fund Rate in two to three steps thus creating with the freshly printed money: Condition 3 where GNP = low, Inflation = low, Fed Fund Rate = low.

After a few months delay for the money to find its way into the pockets of consumers and corporations, the GNP starts to rise and becomes, in our terminology, OK. We have thus arrived at Condition 12 (GNP = OK, Inflation = low, Interest = low). There is only one problem: interest rates are still low, thus easy credit conditions prevail. The economy is now going through Condition 15 (GNP = OK, Inflation = OK, Interest = low). Everybody is convinced now that the bad times are over once and for all. Spot shortages of goods and skilled labor start to show up resulting in a rise of Inflation. Everybody is happy; the economy moves forward through Condition 18 until it reaches the ultimate overheated Condition 27 (GNP = High, Inflation = High, Interest = Low). Buy now — pay later.

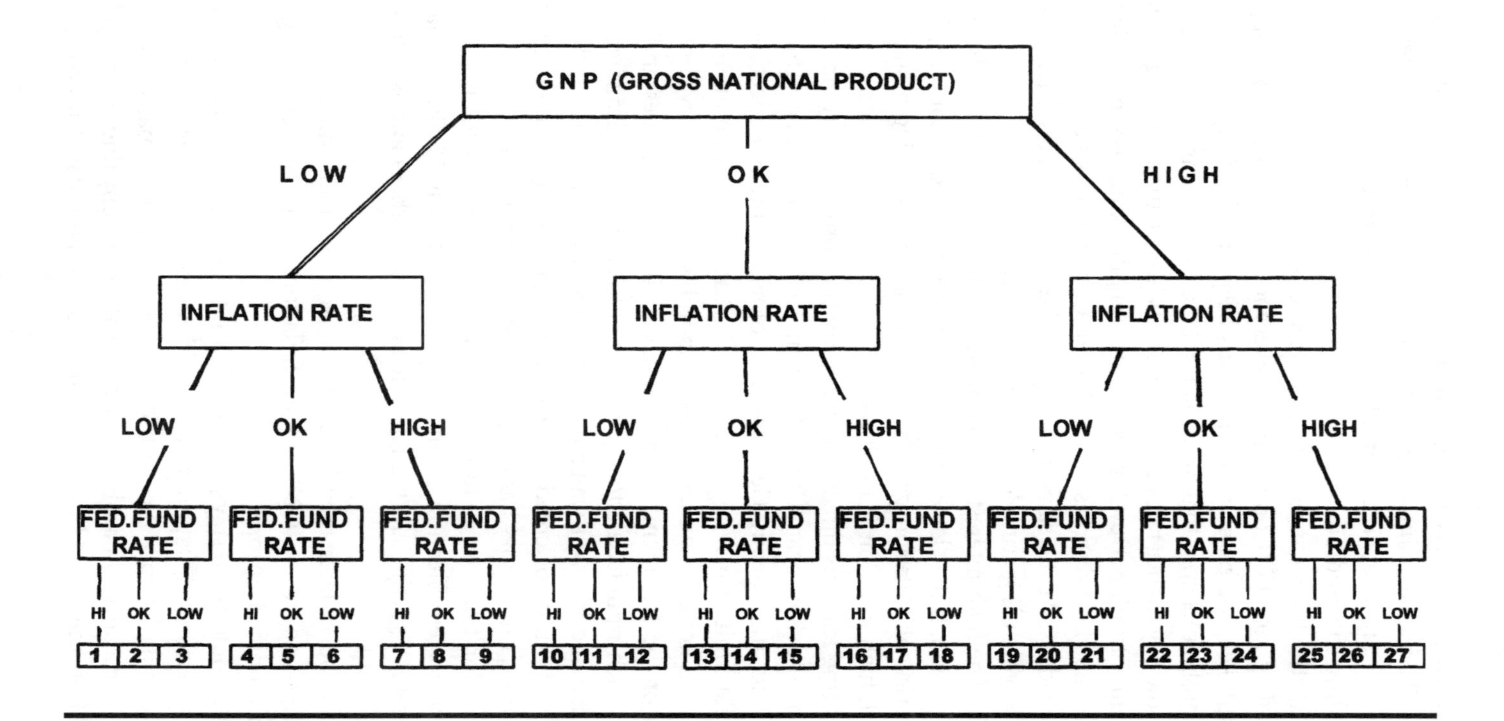

Figure 7.1 Fuzzy Logic Path — Economy

The day of reckoning has arrived. The Chairman of the Federal Reserve Bank doesn't like the latest inflation figures and finally decides to tighten the money supply, i.e., raise the Fed Fund Rate. As a result, interest rates are now going up leading to Conditions 26 and 25.

Meanwhile, the higher interest rates have the desired effect. GNP slows down from a High to OK (Condition 13). Everybody still seems to be happy, consumers still keep on buying, governments keep on spending.

Another problem: banks that have made all these easy loans at low interest rates, must now pay higher rates to depositors. With inflation coming down, real estate values tumble. Developers go into default. Banks go into default or must build up their reserves to stay liquid. Corporations have problems servicing their debt. At the same time, the higher interest rates filter down into the GNP which is getting lower (Condition 4).

Poor Feds. The high interest rates combined with a low GNP has them in a catch-22 situation. If they lower the interest rates to prime the economy, they will run the risk of Inflation getting out of hand thus creating a stagnant economy with high inflation (Condition 9).

The foregoing leads us directly to the Euro, the common currency to be introduced in the European Union (EU) in 1999. Can anybody imagine France, England, Italy, Spain, Germany, and others in the EU, meeting the Maastrich debt limit of maximum 2.5% of GNP at a time when unemployment rates are at double digit levels? We cannot. It's a nice objective but let's face it, each of the EU member nations has their own priorities and will cooperate in such mundane efforts only if it is to their own advantage. We can expect that the EU will water down this requirement considerably or allow nations all kinds of creative bookkeeping to stay within these limits. It's a noble effort, but history has taught us that national interests will always come first. We don't have to go too far back when we look at what happened after the German reunification. In the late 1980s and early 1990s, Germany held its interest rates artificially high despite urgings by other nations to lower rates.

Now, a few years later they have 12% unemployment and a stagnant economy, i.e., Condition 4 (GNP = Low, Inflation = OK, Interest = High). So much for a unified Europe.

An option would be to stay put with high interest rates (as Germany has done) and hope for a deflationary and mild recessionary scenario, the so-called "soft landing." A nice theory, but one that very seldom pans out.

We now have a fairly broad-based idea how the economy undergoes cyclic changes and how the Federal Reserve Bank tries to manipulate the economy.

We have shown how the TOO HIGH-OK-TOO LOW concept can be applied to the economy. We will apply the same concept when we make investment decisions. This technique is the foundation of our investment strategy.

Economists, academia, and financial experts might have some objections to this kind of approach to very complex questions. Some might dismiss our method as being too simplistic perhaps even naive. Our answer: it is not our intention to be economists; all we want to do is to get a good read of the present state of the economy. The Fuzzy Logic approach will do for us quite well. At least it will allow us to make our own independent assessment of the situation without having to listen to the rosy pictures painted by government leaders and assorted experts. We take a look at the present condition and from this assessment we will make our decisions how and where we will invest our money.

Relying on only three indicators to assess the current state of the economy is indeed simplistic. There are a lot of other forces at work that influence the economy, but adding more indicators to our analysis would, for our purpose, not produce a much better assessment. Discount rates, unemployment, trade deficits or factory inventory figures might be of value for others, but not for our purpose. We stand behind our Fuzzy Logic approach. Just imagine: if we would add one more indicator in our analysis, there would be $4 \times 4 \times 4 = 64$ basic conditions and 5 variables would deliver 3125 different conditions we would have to deal with. I think we have made our point.

The reader might have noticed that so far no mention has been made of how our six types (from Chapter 4) of investments react to these 27 basic economic conditions. There is no doubt that the three variables we use for the economic assessment do move the market, be that stocks, bonds, or foreign investments. This omission was done on purpose because each type of investment has its own indicators. In other words, for stock investments we use different indicators and the same applies for bond or foreign investments as we will see in Chapters 8, 9, and 10. Recognition of the current state of the economy is only the first step in our method of analysis. What investment action we will take for any condition in the stock, bond, foreign, or money market will be described at the conclusion of these chapters. There, as part of our Fuzzy Logic concept, we will present tables that show the action keys we use for any of the described conditions. These keys are an intregal part of our strategy.

Let's explore now the indicators that relate directly to the different types of investments.

8 Equity Fund Indicators

Buy low, sell high, the dream strategy of every investor. The problem is that everyone has his own ideas of what is low and what is high. The equity market is alive because there is always someone who thinks a given stock is not worth holding anymore so he sells it to someone who thinks the stock is a good buy. Without such direct opposing viewpoints there wouldn't be a market. When there are more sellers than buyers, the sellers must be satisfied with a little less, thus the market goes down. Conversely, when there are more buyers than sellers, market prices will be bid up to higher levels. It follows the law of supply and demand.

Technicians insist they can predict which way the market or a given stock is going by looking at geometric forms of lines and curves, such head-and-shoulder, double top (M), or double bottom (W) patterns. Fundamentalists, as we have learned, have their own indicators and so do the contrary investors, and the tea leaf readers. Sometimes their predictions prove later to have been correct, but more often are not. Predictions can be correct for the short term, wrong for the long term, and vice-versa.

Brokerage houses are trying to make the small investor believe that in order to make money, one has to be invested at all times and to a substantial degree in the stock market. They use the legitimate argument that a buy-and-hold strategy for the Dow Jones Industrials has historically produced a higher average annual return than bonds or other fixed income investments. There is, however, a dark side to this argument. Reducing, for example, the stock portion of the portfolio from 80% to 70% in difficult market conditions, in our opinion, doesn't classify as sound investment advice. Neither does the notion that a paper loss is not a real loss unless the investment is sold, based on the false assumption that the market will "always" come back to make up the losses.

"Always" could mean years of waiting for the investment to break even. We're not investors who want to be satisfied to break even; we are investors because we want to make a decent profit and have financial security in the future. With our strategy, we do not want to be in equity funds during a bear market — it is as simple as that. Let's once and for all get this myth out of the way. There is no reasonable justification for an investor to have most of his assets in stock funds at all times because, as they say, equity investments are the only investments that over time will produce the best returns.

8.1 The Efficient Market Theory

Lately, the theory of an Efficient Market has come under question from several market experts, including the academic community. This with full justification because there are many legitimate counterarguments speaking against the validity of this theory.

In an Efficient Market, all pertinent information related to a given equity is supposed to be reflected in the price of the stock. Earnings, growth rates, interest rates for the near term future (6-8 month) as well as investors' sentiments would have to be already included in the price of a stock if the theory holds true. In other words, the technical and fundamental analysis reports issued by market analysts for a given stock must therefore be mostly accurate. Investors in individual stocks place a lot of value on such reports and in large part use these to guide them in their buy, hold, or sell decision. But, we know too well how quickly and how often analysts can change their opinions about any given company's future performance and how these reports can move the price of an equity within hours. Is this the "efficiency" the experts are talking about and the investor is looking for?

In the author's opinion, there has never been and will never be an "efficient market." Here are our arguments

1. ***Future interest rates move the price of a market.*** But, decisions and intentions by the central bank related to future interest rate changes are kept under a shroud of secrecy. One never hears a straightforward answer from the chairman of the Federal Reserve Bank such as: when the GNP, unemployment, or inflation rate changes by such-and-such a percentage, we will change the interest rate by so many basis points

(100 points = 1%). As long the public doesn't know what interest rates will be in the next six months, there will be no "Efficient" Market.

2. ***Impending mergers, acquisitions, or restructuring of a company will affect stock prices.*** When insiders know of such impending changes but the public is being left in the dark, insiders have a leg up on the private investors which doesn't speak well for efficiency. Sure, there are insider trading laws to prevent such abuses, but misuse of such information happens all too often, especially in overseas markets. There cannot be an efficient market when certain privileged individuals know more about a company's future than the general public.
3. ***Psychological factors such as investor's sentiments and herd instinct can move markets.*** Markets often reach under- or overvalued levels because of investor's sentiments that can move from fear to greed. Investors sell when the market has dropped or buy when it has already significantly increased — the bandwagon effect that is really not synonymous to an Efficient Market.
4. ***As long as market gurus can move the market with a few short sentences of wisdom*** and investors listen to these fortune tellers, there cannot be an Efficient Market.
5. ***There cannot be an Efficient Market*** as long as investors listen to politicians and believe their words when they proclaim the nation's economy is humming along just fine when, in fact, the country is on the brink of insolvency. Furthermore, as long as governments are willing to "rescue" large companies in financial difficulties and, on the other hand, let thousands of small companies, mom-and-pop shops in much lesser difficulties go under, there can be no market efficiency.
6. ***Liquidity in the marketplace moves stock prices.*** In a raging bull market with plenty of liquidity, so-called deadbeat equities are being bought at unreasonably high prices. Conversely, lack of liquidity in the market can trigger severe market contractions, solid and good equities are being sold where, in fact, they should be held because the long term fundamentals for the company still look good.

Some readers might question the author's definition of an "Efficient Market" raising the argument that in an Efficient Market all the current available information, be that correct or just taken as assumptions, is instantaneously reflected in the stock prices. They might argue that when a CEO resigns and the stock changes significantly within minutes after the announcement, this represents an efficient market. We don't agree because *efficiency,* in our opinion, is synonym to *competent, apt, and qualified,* traits often lacking in the markets.

We *buy* a particular type of investment when conditions are favorable to produce a profit, *hold* that investment as long the profits materialize and, *sell* when conditions turn unfavorable *provided the new overall portfolio risk meets our own personal risk level.*

8.2 Current Stock Market Conditions

Again, as with our model for the economy, our strategy doesn't include any forecasting techniques. All we want is a read of current stock market conditions. Because we do not invest in individual stocks, our work is made considerably easier. All we have to do is look at the stock market index that mirrors the equity market of the country where we want to invest our money. For an American investor this would be the Dow Jones Industrial index (DJ-IND). This is the most widely followed and known index in the U.S. Although mutual fund performance is more often compared to the S&P 500 index, we use the DJ-IND for a specific reason. The DJ-IND index is made up of 30 of the best known large U.S. capitalization stocks which are clearly favored by foreign investors. The U.S. stock market has ceased to be a market governed by domestic investors alone. Foreign investor buying or selling, by itself, can move the U.S. equity markets. By following the DJ-IND index, we have a read of how American and foreign investors valuate the U.S. market.

We look at the stock market index knowing with ironclad certainty that it will neither go up or down forever and never stays the same for any prolonged period of time. Like the economy, but not in lock-step, the stock market will follow a cyclic pattern. Each cycle will be different in magnitude and duration and thus has its own fingerprints. Again, nobody knows how long the waves in each of the cycles will last, nor does anybody know how high or low the waves will be.

8.3 The Economy as an Indicator for Stock Market Conditions

The first indicator we look at is the previously discussed economic condition because the direction of the economy has a profound influence on the equity market. Described in detail in the previous chapter, we have an economy that expands (high), contracts (low), or an economy that is more or less stable (OK). We take the number for the economic condition from Figure 7.1 and enter it into Figure 8.1. Later we will add the remaining data.

Many techniques can be used to determine if the market is overvalued, fairly priced, or undervalued. Formulas have been developed by many to answer the question at what price level the stock market realistically should be. This type of analysis calls for finding the "intrinsic" value of the market or of individual stocks using prevailing or projected earnings and interest rate conditions. (We will discuss this in greater detail in 15.2). We, too, use such an indicator to tell us if the market value is currently too high, OK, or too low. Simply stated, we need such an indicator because we do not want to buy something whose price is too high but rather buy when the price is a bargain. What applies to the purchase of a car applies equally to the purchase of a mutual fund. We are willing to buy when the price is low or reasonable, but refuse to buy when it is too high.

8.4 The Price-to-Earnings Ratio

This then is our second indicator, the Price-to-Earnings ratio (P/E) of the market as a whole. For U.S. investors this would be the P/E of the Dow Jones Industrial averages. A German would use the DAX and a British investor the FT-100 for this purpose.

Earnings generated by a corporation is one of the prime factors used by investors for market and individual stock valuation. When earnings rise, investors tend to drive the market up. Likewise, when earnings fall or are reported lower than what the market expected, prices will fall. Again, there are other indicators that could be used for the same purpose of valuation such as Price/Book, Dividend Yield, Relative Strength, etc. We stay with the P/E ratio for our analysis because we want to keep the number of indicators down to a minimum in order to not cloud the overall important picture.

A word of caution: a low P/E ratio doesn't necessarily indicate the market is undervalued and a high ratio that the market is overvalued. High P/E ratios quite often occur during the tail end of a recession. In such times, a high P/E ratio is usually not the result of overvaluation but because of persistent low earnings of corporations although the market prices have already been driven up by investor anticipation of an economic rebound. Strong evidence exists that stock prices don't reflect current conditions but what investors expect it to be six or eight months into the future. This supports our contention of not relying on one indicator alone, such as the P/E ratio, to make an investment decision. Other pertinent indicators have to be considered as well to make sure the decision is sound and justifiable.

There are also some market experts who have statistically proven that investors are willing to buy equities when the market P/E ratio is relative high, say over 19, at times when inflation and interest rates are low and the nation's economy is humming along fine. Conversely, they say investors tend to buy lower market P/Es (less than 10) during recessionary, high interest, and high inflation times. We do not directly differentiate "adjustable" P/E ratios in this manner in our strategy. The economy, interest, and inflation rates are separate entities we watch and are given due consideration in our Fuzzy Logic approach.

In our strategy, we consider

P/E Ratio less than 12 = Low
P/E Ratio 12 to 18 = OK
P/E Ratio higher than 18 = High

After a severe recession with unusual low earnings by most companies, all P/E ratios are considered OK for our purpose. It doesn't mean they are, but are earmarked as such in our evaluation procedure.

8.5 The Gap Indicator

The third indicator we look at to determine the state of the stock market is the Gap, i.e., the difference between the current market price and the weighted moving average (WMA). Moving averages have long been used by investors to smooth out daily and weekly volatilities prevalent in the equity markets. At any given time, the DJ-IND index could be above or below this moving average. The 200-day moving average is perhaps the most widely followed trend line. To calculate this average, the last 200 days DJ-IND prices are added and the sum divided by 200. Another popular average is the 50-day moving average. Both these averages are used by a multitude of mutual fund investors for buy and sell signals. They buy when the DJ-IND prices move above and sell when the price moves below these moving average lines.

The problem with these simple moving averages is that equal weight is being given to each individual price be that one of yesterday, last week, or a few months ago. The fact is that some of the data is history that has most probably lost its significance on the current market trend. Some mathematicians and investors have recognized this fact and developed sophisticated weighted exponential moving averages that assign more weight to most recent prices and less weight to older data.

I, too, have developed such a weighted moving average suitable for our purpose applicable to equity mutual fund investments. The mathematics involved are not much different than perhaps thousands of similar known weighted moving averages except I apply different weights and different time spans to calculate our weighted moving average (WMA). What does set us apart from the others is that we do not use our WMA for sell, hold, or buy signals. The calculated WMA is only of secondary value because its sole purpose for us is to calculate the gap, the difference between the current market index level and the weighted moving average trend line.

Critics of weighted moving averages point out that the 200-day average often gives a late signal when the market has already turned to a significant degree. This was the case in October 1987 when the 200-day average didn't trigger a sell signal until three days before the market crash. It caught many weekend market watchers too late to save their equity portfolios from a precarious drop. The deficiencies of the 50-day moving average is that it often generates false signals and causes whipsaws wherein the market turns around again soon after investors have executed a sell or buy order. We do not want to give the impression these moving averages are useless. They have served many investors well sometimes but rarely all the time. The error as we see it is not the way these averages are calculated, but the often false assumption made by investors that the moving average is the sole indicator to be used for buy, sell, and hold signals. It would be nice if things would be that easy, but they are not as any investor can attest after he has followed the markets for a few years.

We have learned before: single mindedness around one particular indicator alone for investment decisions can lead to poor portfolio performance. Becoming a consistent successful investor entails a little more work and effort than to just watch one single indicator.

We look at our gap indicator from a quite different viewpoint, thus, I will try to explain this concept in more detail because it is important to the reader if he wants to become familiar with this investment philosophy.

In our concept for evaluating the current market trend we use the following basic thoughts:

1. The market will always be in an up, a down, or sideways trend. It will oscillate through time. Big waves, small waves, short and long waves. We only know what the past and current waves looked like but will not know the character of the next wave. The equity market is not repetitious like the ocean where the sixth and seventh waves are usually the largest. The market can deliver a whopper of a wave at any given time it pleases. Remember, we are not surfers, we are investors.
2. Weighted moving averages give only an indication of the market trend. Using a moving average will only serve to tell if the market is moving with or against that trend.
3. Periodically, market prices tend to significantly move above or below the moving average. This is a direct reflection of investor's sentiments and their collective instinct to stampede in or out of the market at the same time. These excessive moves away from the moving average, in due time, will correct themselves. In other words, when the gap between the prices and the moving average becomes too large, there soon will be a countermove that will return prices closer to the moving average. In our technique, we take advantage of these excessive market moves. Our way classifies as contrarian thinking because we view such large gaps as an indication that the market has gotten ahead of itself. Large positive gaps are danger signals for us that the market is speculatively overheated. Conversely, a large negative gap indicates too much bearishness among investors.
4. Good opportunities to sell and buy equity funds are during periods of excessive gaps between the current market price and the weighted moving average.

The author's own weighted moving average has served his investment strategy well when combined with the other mentioned indicators. Most important, the Fuzzy Logic method has allowed him to be out of equity funds well before the 1987 market crash and the 1990 and 1998 corrections. It has produced returns from the equity portion of his portfolio that contributed significantly toward the steady and consistent growth of his portfolio. Consistency is an important attribute of any investment strategy. The results are shown in Figures 8.2 to 8.12, and Tables 8.1 to 8.11 covering the period from 1986 to 1998. These graphs and tables show the prices and moving averages for the Dow Jones Industrials. The buy and sell opportunities are recognized with the marks ($) and (L), respectively, during periods when the gaps were excessively large to the up or down side relative to the moving average.

The weighted moving average reflects our perception of the importance the gap plays as a current market trend indicator. The gap is expressed as a percentage by which the current market index deviates from the current WMA. It is important to mention that we calculate the weighted moving average and the gap on single weekly closing prices.

$$\text{WMA} = 0.128 * \text{current DJ-IND} + 0.872 * \text{previous week WMA} \quad 8.5.1$$

$$\text{GAP} = \left(\left(\text{current DJ-IND}/\text{current WMA}\right) - 1\right) * 100 \quad 8.5.2$$

The thus determined buy and sell opportunities are viewed in context with the other indicators and are specifically designed for equity mutual fund investments. The ultimate decision to buy or sell equity fund assets are made only in conjunction with the results of our risk and asset allocation analysis. For example, when our equity fund Fuzzy Logic condition indicates a buying opportunity, a purchase is executed only when this action meets our risk and asset allocation goals. The same applies to the selling of equity funds or, for that matter, for any of the other fund types as well. Buy, sell, or hold opportunities are signalled by the Fuzzy Logic Keys discussed in Chapter 12.

Unfortunately, to calculate the weighted moving average and the gap by hand is cumbersome but it can be quickly done and plotted by means of a computer. In our Fuzzy Logic assessment we consider

a gap of more than + 5.0% = High
a gap between – 3.0 and + 5.0% = OK
a gap lower than – 3.0% = Low

8.6 The Fuzzy Logic Path — Equity Markets

In summary, to assess the current state of the stock market we look at three key indicators and follow along the path on Figure 8.1 to arrive at the current condition of the stock market.

1. Economy: expanding, contracting, or OK?
2. P/E ratio: too high, too low, or OK?
3. Gap: too far above(High), too far below (Low), or OK compared to WMA?

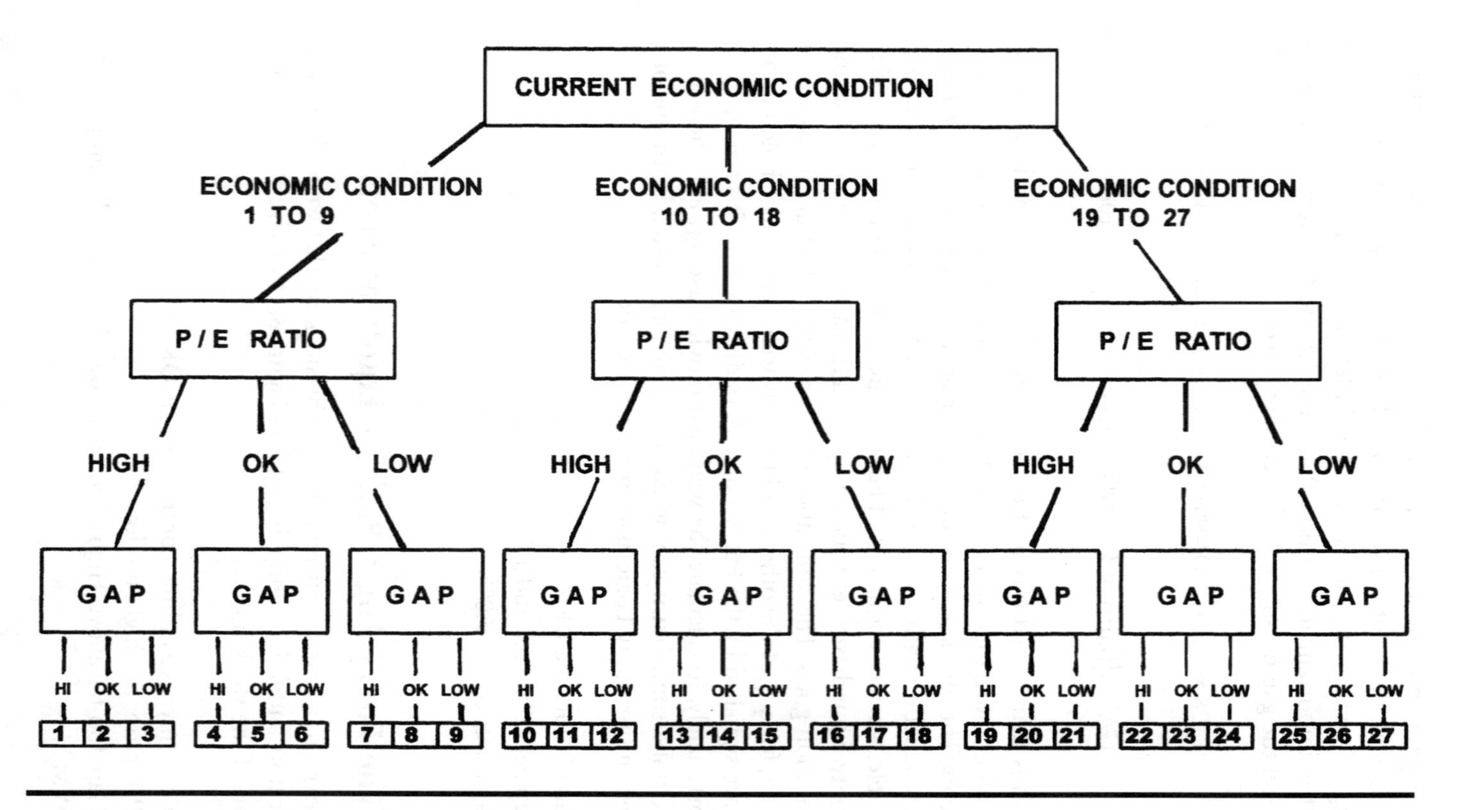

Figure 8.1 Fuzzy Logic Path — Equity Fund Investments

For example, at the end of 1997 the following conditions prevailed: economy Condition 14, P/E ratio too high, gap OK giving us the stock market condition number **11**. This is the number we have to note down because it will be the key for us to determine if good opportunities exist to either buy or sell equity investments.

We thus again have a total of $3 \times 3 \times 3 = 27$ basic stock market conditions shown graphically in Figure 8.1.

No claim is made that this is the ultimate panacea and the Holy Grail of equity mutual fund timing. So far this method has produced remarkably good results for the past few years, but only time will tell how accurate this method is to identify buy and sell opportunities for equity mutual funds in the future. We are realists and accept the fact that no mathematical model will ever be devised that can get an investor into stock mutual funds at the precise bottom and out at the exact top of the market accurately and error-free. And so it should be. With this described method we lock in the profits and don't worry about making a few extra percent by continuing to ride a trend when our own risk profile doesn't allow us to do so. There is no need to expose our investment portfolio to undue risk by staying in equity mutual funds at all times. After all, there are five other investment types where our money could possibly produce a profit, too. One should never be married to stock mutual funds alone; they are not the only game in town.

As most investors that rode the wild gyrations of the market in the late 1980s and 1990s can attest prematurely taking profits is not only prudent but also considered to be street smart. Everybody can look at a chart depicting the past and render a judgement on what should or should not have been done. Making a decision at the time when the future is unknown is much more difficult. The reader can blank off part of any graph presented in this chapter and ask himself the question how he would have assessed the market performance at that particular time and if he would have made an adjustment in his equity holdings. I think I made the point.

Let's assume our Fuzzy Logic indicators signal an opportune time to purchase equity mutual funds. The legitimate question could then be raised if equity funds indeed should be purchased each time such an indication is given. The answer is yes provided, and this is important, our asset allocation and risk analysis have shown these additional purchases would not upset our initial goals and risk levels for our portfolio as a whole (as discussed in Chapters 5 and 6). On the other hand, if the portfolio already contains the necessary amount of equity fund investments, we would not make additional investments into these funds even when our indicators show a buy opportunity. This concept is an important aspect of our investment style and therefore emphasized as follows:

Axiom VII

Any portfolio adjustments must be made with the objectives of the portfolio risk matching closely the individual investor's own risk level.

As the Dow Jones Industrials graphs clearly show, the gap indicator by itself cannot be used for a buy or sell decision. However, it is a powerful tool when used in conjunction with the other indicators to make a meaningful assessment of current opportunities for equity fund investments. At times, this might appear to be a conservative approach, but it suits us fine for we are primarily interested in consistent returns and do not hesitate to exit the equity market when it appears to get too volatile for our comfort.

In the following graphs we have marked the selling opportunities with "**$**" and the buying opportunities with "**L**" when the gap was excessively high or low.

How important it is not to use the gap indicator alone for equity fund buy and sell decisions can be demonstrated by the following:

If an investor would have rigidly sold on "**$**" signals and bought on "**L**" signals, his total gain (simple not compounded) would have been 153.3 percent over the period from 1986 to 1996, assuming he would have invested the proceeds in a money market account in the periods when he was out of the stock market. Another investor, if he would have bought in 1986 and held for the entire period until the end of 1996 would have achieved a gain of 223 percent. What the gap indicator however did was to flash warning signals of overvalued markets weeks ahead of the market shocks in 1987, 1990, 1997, and 1998. This clearly underscores the rule we have repeatedly mentioned earlier, namely,

---

Axiom VIII

Do not attempt to time the markets by switching in and out of funds whenever a single indicator flashes a buy or sell signal.

Table 8.1 DJ-IND 1987 Gap from WMA

Week	DJ-IND	WMA	High Gap	Low Gap	GAP	Week	DJ-IND	WMA	High Gap	Low Gap	GAP
1	1895	1887			4.5	27	2437	2329			4.6
2	2005	1902	$		5.4	28	2456	2345			4.7
3	2076	1924	$		7.9	29	2510	2366	$		6.1
4	2101	1947	$		7.9	30	2485	2382			4.3
5	2158	1974	$		9.3	31	2572	2406	$		6.9
6	2186	2001	$		9.2	32	2592	2430	$		6.7
7	2183	2024	$		7.8	33	2685	2462	$		9.0
8	2235	2051	$		8.9	34	2709	2494	$		8.6
9	2223	2073	$		7.2	35	2639	2513	$		5.0
10	2280	2100	$		8.6	36	2561	2519			1.7
11	2258	2120	$		6.5	37	2608	2530			3.1
12	2333	2147	$		8.6	38	2524	2529			–0.2
13	2335	2171	$		7.5	39	2570	2535			1.4
14	2390	2199	$		8.7	40	2640	2548			3.6
15	2338	2217	$		5.5	41	2482	2540			–2.3
16	2275	2224			2.3	42	2246	2502		L	–10.2
17	2235	2226			0.4	43	1950	2431		L	–19.8
18	2280	2233			2.1	44	1993	2375		L	–16.1
19	2322	2244			3.5	45	1959	2322		L	–15.6
20	2272	2248			1.1	46	1935	2272		L	–14.8
21	2243	2247			–0.2	47	1913	2226		L	–14.1
22	2291	2253			1.7	48	1910	2186		L	–12.6
23	2326	2262			2.8	49	1766	2132		L	–17.2
24	2377	2277			4.4	50	1867	2098		L	–11.0
25	2421	2295	$		5.5	51	1975	2082		L	–5.2
26	2436	2313	$		5.3	52	1938	2048		L	–5.4

Table 8.2 DJ-IND 1988 Gap from WMA

Week	DJ-IND	WMA	High Gap	Low Gap	GAP	Week	DJ-IND	WMA	High Gap	Low Gap	GAP
1	1911	2037		L	–9.3	27	2106	2062			2.1
2	1956	2026		L	–3.5	28	2129	2071			2.8
3	1903	2010		L	–5.3	29	2061	2070			–0.4
4	1958	2004			–2.3	30	2128	2077			2.4
5	1910	1992		L	–4.1	31	2119	2082			1.8
6	1983	1991			–0.4	32	2037	2077			–1.9
7	2015	1994			1.1	33	2016	2069			–2.6
8	2023	1998			1.3	34	2017	2062			–2.2
9	2057	2005			2.6	35	2054	2061			–0.3
10	2034	2009			1.3	36	2068	2062			0.3
11	2087	2019			3.4	37	2098	2067			1.5
12	1978	2014			–1.8	38	2090	2070			1.0
13	1988	2010			–1.1	39	2112	2075			1.8
14	2090	2021			3.4	40	2150	2085			3.1
15	1980	2015			–1.8	41	2133	2091			2.0
16	2015	2015			0.0	42	2183	2103			3.8
17	2032	2017			0.7	43	2149	2109			1.9
18	2007	2016			–0.5	44	2145	2113			1.5
19	1990	2013			–1.1	45	2067	2107			–1.9
20	1952	2005			–2.6	46	2062	2102			–1.9
21	1956	1999			–2.1	47	2075	2098			–1.1
22	2071	2008			3.1	48	2092	2097			–0.3
23	2102	2020			4.1	49	2143	2103			1.9
24	2104	2031			3.6	50	2151	2109			2.0
25	2142	2045			4.7	51	2168	2117			2.4
26	2131	2056			3.6	52	2168	2123			2.1

Table 8.3 DJ-IND 1989 Gap from WMA

Week	DJ-IND	WMA	High Gap	Low Gap	GAP	Week	DJ-IND	WMA	High Gap	Low Gap	GAP
1	2194	2132			2.5	27	2487	2438			2.0
2	2226	2144			3.8	28	2554	2453			4.1
3	2235	2156			3.7	29	2607	2473	$		5.4
4	2322	2177	$		6.7	30	2635	2493	$		5.7
5	2331	2197	$		6.1	31	2653	2514	$		5.5
6	2286	2208			3.5	32	2684	2536	$		5.9
7	2324	2223			4.5	33	2687	2555	$		5.2
8	2245	2226			0.9	34	2732	2578	$		6.0
9	2274	2232			1.9	35	2752	2600	$		5.8
10	2282	2238			1.9	36	2709	2614			3.6
11	2292	2245			2.1	37	2674	2622			2.0
12	2243	2245			–0.1	38	2681	2629			2.0
13	2293	2251			1.9	39	2692	2637			2.1
14	2304	2258			2.0	40	2785	2656			4.9
15	2337	2268			3.0	41	2569	2645			–2.9
16	2409	2286	$		5.4	42	2689	2651			1.4
17	2418	2303			5.0	43	2596	2644			–1.8
18	2381	2313			2.9	44	2629	2642			–0.5
19	2439	2329			4.7	45	2625	2640			–0.6
20	2501	2351	$		6.4	46	2652	2641			0.4
21	2493	2369	$		5.2	47	2675	2646			1.1
22	2517	2388	$		5.4	48	2747	2659			3.3
23	2513	2404			4.5	49	2731	2668			2.4
24	2486	2415			3.0	50	2739	2677			2.3
25	2531	2430			4.2	51	2711	2681			1.1
26	2440	2431			0.4	52	2753	2690			2.3

Table 8.4 DJ-IND 1990 Gap from WMA

Week	DJ-IND	WMA	High Gap	Low Gap	GAP	Week	DJ-IND	WMA	High Gap	Low Gap	GAP
1	2753	2698			3.9	27	2880	2792			3.1
2	2773	2708			2.4	28	2960	2814	$		5.2
3	2689	2705			–0.6	29	2980	2835	$		5.1
4	2677	2702			–0.9	30	2961	2851			3.9
5	2559	2683		L	–4.6	31	2898	2857			1.4
6	2603	2673			–2.6	32	2809	2851			–1.5
7	2648	2670			–0.8	33	2716	2834		L	–4.2
8	2635	2665			–1.1	34	2644	2809		L	–5.9
9	2564	2652		L	–3.3	35	2533	2774		L	–8.7
10	2660	2653			0.2	36	2614	2754		L	–5.1
11	2683	2657			1.0	37	2619	2736		L	–4.3
12	2741	2668			2.7	38	2564	2714		L	–5.5
13	2704	2673			1.2	39	2512	2688		L	–6.6
14	2707	2677			1.1	40	2452	2658		L	–7.8
15	2717	2682			1.3	41	2510	2639		L	–4.9
16	2751	2691			2.2	42	2398	2608		L	–8.1
17	2696	2692			0.2	43	2520	2597			–3.0
18	2645	2686			–1.5	44	2436	2576		L	–5.4
19	2710	2689			0.8	45	2490	2565			–2.9
20	2801	2703			3.6	46	2488	2555			–2.6
21	2819	2718			3.7	47	2550	2555			–0.2
22	2820	2731			3.3	48	2527	2551			–0.9
23	2900	2753	$		5.4	49	2559	2552			0.3
24	2682	2744			–2.2	50	2590	2557			1.3
25	2935	2768	$		6.0	51	2593	2562			1.2
26	2857	2779			2.8	52	2633	2571			2.4

Table 8.5 DJ-IND 1991 Gap from WMA

Week	DJ-IND	WMA	High Gap	Low Gap	GAP	Week	DJ-IND	WMA	High Gap	Low Gap	GAP
1	2633	2579			0.4	27	2932	2917			0.5
2	2566	2577			–0.4	28	2980	2925			1.9
3	2501	2568			–2.6	29	3016	2937			2.7
4	2646	2578			2.7	30	2972	2941			1.0
5	2659	2588			2.7	31	3006	2949			1.9
6	2730	2606			4.8	32	2996	2955			1.4
7	2830	2635	$		7.4	33	2968	2957			0.4
8	2934	2673	$		9.8	34	3040	2968			2.4
9	2889	2701	$		7.0	35	3011	2973			1.3
10	2909	2727	$		6.7	36	2985	2975			0.3
11	2955	2757	$		7.2	37	3019	2980			1.3
12	2948	2781	$		6.0	38	3006	2984			0.7
13	2858	2791			2.4	39	2961	2981			–0.7
14	2913	2807			3.8	40	2983	2981			0.1
15	2896	2818			2.8	41	3077	2993			2.8
16	2920	2831			3.1	42	3004	2995			0.3
17	2965	2848			4.1	43	3056	3003			1.8
18	2912	2856			1.9	44	3045	3008			1.2
19	2938	2867			2.5	45	2943	3000			–1.9
20	2920	2874			1.6	46	2902	2987			–2.9
21	2886	2875			0.4	47	2894	2975			–2.7
22	2913	2880			1.1	48	2886	2964			–2.6
23	3027	2899			4.4	49	2914	2957			–1.5
24	2976	2909			2.3	50	2934	2954			–0.7
25	2965	2916			1.7	51	3101	2973			4.3
26	2906	2915			–0.3	52	3168	2998	$		5.7

Table 8.6 DJ-IND 1992 Gap from WMA

Week	DJ-IND	WMA	High Gap	Low Gap	GAP	Week	DJ-IND	WMA	High Gap	Low Gap	GAP
1	3168	3021			3.9	27	3282	3315			–1.0
2	3201	3044	$		5.2	28	3330	3317			0.4
3	3199	3064			4.4	29	3330	3318			0.3
4	3264	3089	$		5.7	30	3331	3320			0.3
5	3232	3108			4.0	31	3285	3316			–0.9
6	3223	3122			3.2	32	3393	3325			2.0
7	3225	3135			2.9	33	3332	3326			0.2
8	3245	3149			3.0	34	3328	3327			0.0
9	3280	3166			3.6	35	3254	3317			–1.9
10	3267	3179			2.8	36	3267	3311			–1.3
11	3221	3184			1.1	37	3281	3307			–0.8
12	3235	3191			1.4	38	3306	3307			0.0
13	3276	3202			2.3	39	3327	3309			0.5
14	3231	3206			0.8	40	3250	3302			–1.6
15	3249	3211			1.2	41	3200	3289			–2.7
16	3255	3217			1.2	42	3136	3269		L	–4.1
17	3366	3236			4.0	43	3174	3257			–2.5
18	3324	3247			2.4	44	3207	3251			–1.3
19	3336	3258			2.4	45	3226	3247			–0.7
20	3369	3273			2.9	46	3240	3247			–0.2
21	3353	3283			2.1	47	3227	3244			–0.5
22	3386	3296			2.7	48	3282	3249			1.0
23	3396	3309			2.6	49	3304	3256			1.5
24	3398	3320			2.3	50	3313	3263			1.5
25	3354	3325			0.9	51	3326	3271			1.7
26	3285	3320			–1.0	52	3301	3275			0.8

Table 8.7 DJ-IND 1993 Gap from WMA

Week	DJ-IND	WMA	High Gap	Low Gap	GAP	Week	DJ-IND	WMA	High Gap	Low Gap	GAP
1	3301	3280			1.1	27	3483	3465			0.5
2	3251	3276			–0.8	28	3521	3472			1.4
3	3271	3276			–0.1	29	3528	3479			1.4
4	3256	3273			–0.5	30	3546	3488			1.7
5	3310	3278			1.0	31	3539	3494			1.3
6	3442	3299			4.3	32	3560	3503			1.6
7	3392	3311			2.5	33	3569	3511			1.6
8	3322	3312			0.3	34	3615	3525			2.6
9	3370	3320			1.5	35	3640	3539			2.8
10	3404	3330			2.2	36	3633	3551			2.3
11	3427	3343			2.5	37	3621	3560			1.7
12	3471	3359			3.3	38	3613	3567			1.3
13	3439	3369			2.1	39	3545	3564			–0.5
14	3370	3369			0.0	40	3581	3566			0.4
15	3396	3373			0.7	41	3584	3569			0.4
16	3478	3386			2.7	42	3629	3576			1.5
17	3413	3390			0.7	43	3649	3586			1.8
18	3427	3395			1.0	44	3680	3598			2.3
19	3437	3400			1.1	45	3643	3604			1.1
20	3443	3405			1.1	46	3684	3614			1.9
21	3492	3417			2.2	47	3684	3623			1.7
22	3527	3431			2.8	48	3704	3633			1.9
23	3545	3445			2.9	49	3740	3647			2.6
24	3506	3453			1.5	50	3751	3660			2.5
25	3494	3458			1.0	51	3757	3673			2.3
26	3490	3462			0.8	52	3754	3683			1.9

Table 8.8 DJ-IND 1994 Gap WMA

Week	*DJ-IND*	*WMA*	*High Gap*	*Low Gap*	*GAP*	*Week*	*DJ-IND*	*WMA*	*High Gap*	*Low Gap*	*GAP*
1	3754	3696			3.1	27	3646	3722			–2.0
2	3820	3712			2.9	28	3709	3720			–0.3
3	3867	3732			3.6	29	3753	3724			0.8
4	3914	3755			4.2	30	3735	3726			0.2
5	3945	3780			4.4	31	3764	3731			0.9
6	3871	3791			2.1	32	3747	3733			0.4
7	3894	3804			2.4	33	3768	3737			0.8
8	3887	3815			1.9	34	3755	3740			0.4
9	3832	3817			0.4	35	3881	3758			3.3
10	3832	3819			0.3	36	3885	3774			2.9
11	3862	3825			1.0	37	3874	3787			2.3
12	3895	3834			1.6	38	3933	3805			3.4
13	3774	3826			–1.4	39	3831	3809			0.6
14	3635	3802		L	–4.4	40	3843	3813			0.8
15	3674	3785			–2.9	41	3797	3811			–0.4
16	3661	3769			–2.9	42	3910	3824			2.3
17	3648	3754			–2.8	43	3891	3832			1.5
18	3681	3744			–1.7	44	3930	3845			2.2
19	3669	3735			–1.8	45	3807	3840			–0.9
20	3659	3725			–1.8	46	3801	3835			–0.9
21	3766	3730			1.0	47	3815	3832			–0.5
22	3757	3734			0.6	48	3708	3817			–2.8
23	3772	3739			0.9	49	3745	3807			–1.6
24	3773	3743			0.8	50	3691	3792			–2.7
25	3776	3747			0.8	51	3807	3794			0.3
26	3636	3733			–2.6	52	3834	3799			0.9

Table 8.9 DJ-IND 1995 Gap from WMA

Week	*DJ-IND*	*WMA*	*High Gap*	*Low Gap*	*GAP*	*Week*	*DJ-IND*	*WMA*	*High Gap*	*Low Gap*	*GAP*
1	3834	3808			0.5	27	4556	4363			4.4
2	3867	3815			1.4	28	4702	4407	$		6.7
3	3908	3827			2.1	29	4708	4445	$		5.9
4	3869	3833			0.9	30	4641	4470			3.8
5	3857	3836			0.6	31	4715	4502			4.7
6	3928	3848			2.1	32	4683	4525			3.5
7	3939	3859			2.1	33	4660	4542			2.6
8	3953	3871			2.1	34	4640	4555			1.9
9	4011	3889			3.1	35	4590	4559			0.7
10	3989	3902			2.2	36	4640	4569			1.5
11	4035	3919			3.0	37	4730	4590			3.0
12	4073	3939			3.4	38	4840	4622			4.7
13	4138	3964			4.4	39	4850	4651			4.3
14	4157	3989			4.2	40	4860	4678			3.9
15	4192	4015			4.4	41	4850	4700			3.2
16	4208	4040			4.2	42	4840	4718			2.6
17	4270	4069			4.9	43	4810	4730			1.7
18	4321	4101	$		5.4	44	4780	4736			0.9
19	4343	4132	$		5.1	45	4735	4736			0.0
20	4430	4170	$		6.2	46	4718	4734			–0.3
21	4341	4192			3.5	47	4795	4742			1.1
22	4369	4215			3.7	48	4840	4754			1.8
23	4444	4244			4.7	49	5008	4787			4.6
24	4423	4267			3.7	50	5035	4818			4.5
25	4510	4298			4.9	51	5165	4863	$		6.2
26	4585	4335	$		5.8	52	5055	4887			3.4

Table 8.10 DJ-IND 1996 Gap from WMA

Week	*DJ-IND*	*WMA*	*High Gap*	*Low Gap*	*GAP*	*Week*	*DJ-IND*	*WMA*	*High Gap*	*Low Gap*	*GAP*
1	5055	4923	$		5.9	27	5654	5563			1.6
2	5135	4945			3.8	28	5588	5560			0.5
3	5010	4948			1.2	29	5510	5548			–0.7
4	5168	4972			4.0	30	5349	5517		L	–3.1
5	5250	5002			5.0	31	5320	5487		L	–3.0
6	5375	5045	$		6.5	32	5460	5478			–0.3
7	5545	5104	$		8.6	33	5530	5479			0.9
8	5505	5150	$		6.9	34	5605	5490			2.1
9	5630	5206	$		8.1	35	5580	5496			1.5
10	5490	5238			4.8	36	5505	5491			0.2
11	5465	5261			3.9	37	5549	5493			1.0
12	5580	5297	$		5.3	38	5620	5504			2.1
13	5630	5334	$		5.5	39	5828	5540	$		5.2
14	5587	5361			4.2	40	5888	5579	$		5.5
15	5680	5397	$		5.2	41	5992	5626	$		6.5
16	5535	5409			2.3	42	6091	5680	$		7.2
17	5516	5417			1.8	43	6112	5730	$		6.7
18	5565	5431			2.5	44	6145	5777	$		6.4
19	5480	5432			0.9	45	6223	5828	$		6.8
20	5515	5437			1.4	46	6348	5889	$		7.8
21	5690	5464			4.1	47	6523	5964	$		9.4
22	5774	5498	$		5.0	48	6354	6008	$		5.8
23	5640	5511			2.3	49	6304	6040			4.4
24	5698	5529			3.1	50	6450	6087	$		6.0
25	5660	5540			2.2	51	6560	6141	$		6.8
26	5703	5556			2.7	52	6445	6174			4.4

Table 8.11 DJ-IND 1997 Gap from WMA

Week	DJ-IND	WMA	High Gap	Low Gap	GAP	Week	DJ-IND	WMA	High Gap	Low Gap	GAP
1	6840	6174	$		10.8	27	7899	7288	$		8.4
2	6699	6235	$		7.4	28	7920	7362	$		7.6
3	6805	6302	$		8.0	29	8150	7455	$		9.3
4	6855	6366	$		7.7	30	8200	7543	$		8.7
5	6871	6425	$		6.9	31	8025	7597	$		5.6
6	7001	6492	$		7.8	32	7690	7602			1.2
7	6925	6541	$		5.9	33	7895	7632			3.5
8	6890	6579			4.7	34	7620	7622			0.0
9	7002	6627	$		5.7	35	7825	7641			2.4
10	6940	6660			4.2	36	7740	7646			1.2
11	6800	6671			1.9	37	7920	7673			3.2
12	6750	6675			1.1	38	7930	7698			3.0
13	6530	6650			–1.8	39	8030	7733			3.8
14	6555	6631			–1.1	40	8035	7764			3.5
15	6398	6594			–3.0	41	7860	7769			1.2
16	6700	6601			1.5	42	7730	7756			–0.3
17	6740	6612			1.9	43	7450	7709		L	–3.4
18	7070	6664	$		6.1	44	7595	7687			–1.2
19	7130	6717	$		6.1	45	7580	7665			–1.1
20	7190	6771	$		6.2	46	7880	7685			2.5
21	7350	6838	$		7.5	47	7820	7695			1.6
22	7330	6895	$		6.3	48	8209	7753	$		5.9
23	7425	6956	$		6.7	49	7820	7754			0.9
24	7790	7055	$		10.4	50	7770	7748			0.3
25	7799	7144	$		9.2	51	7680	7732			–0.7
26	7695	7207	$		6.8	52	7901	7746			2.0

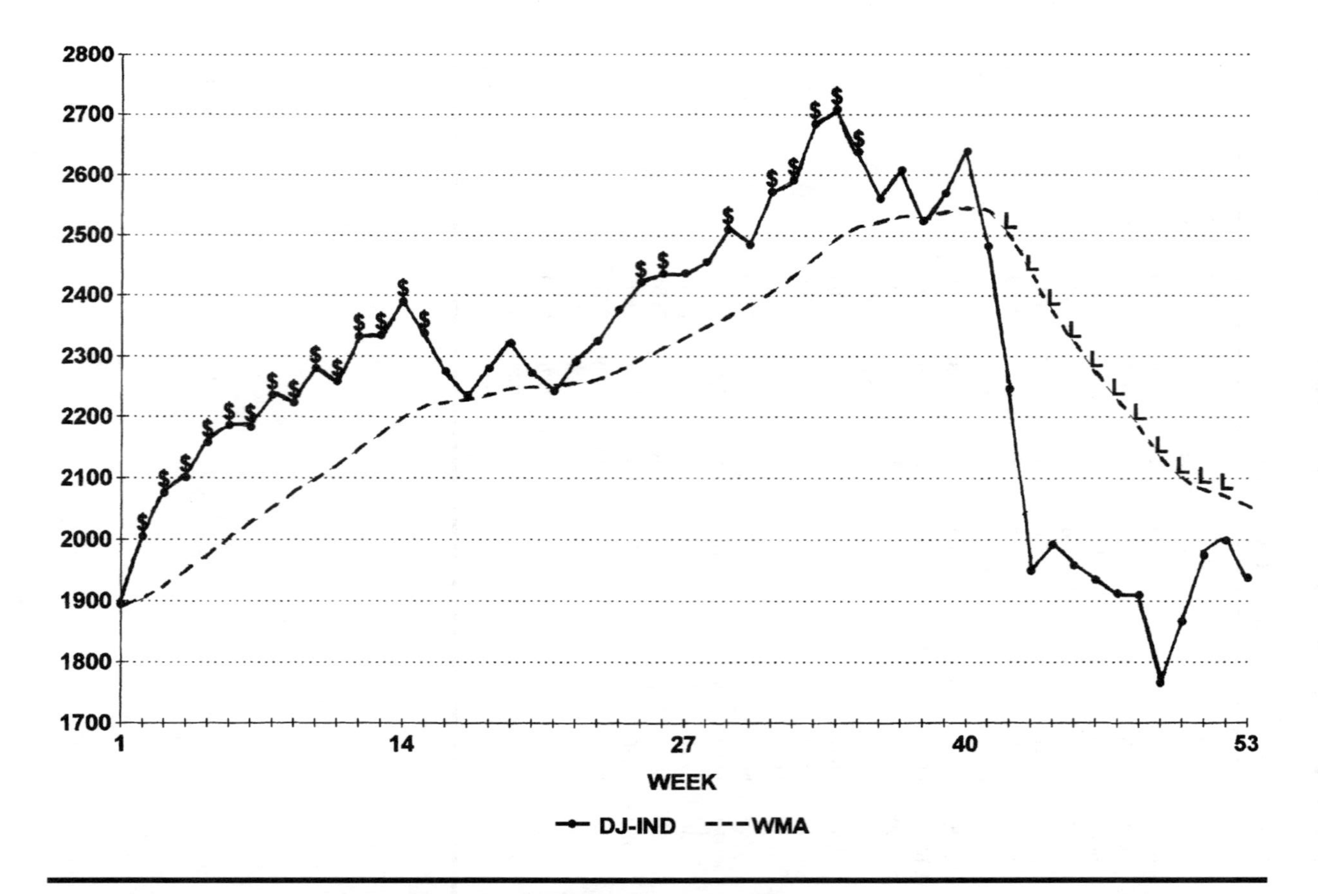

Figure 8.2 DJ-Industrials 1987 GAP from Weighted Moving Average

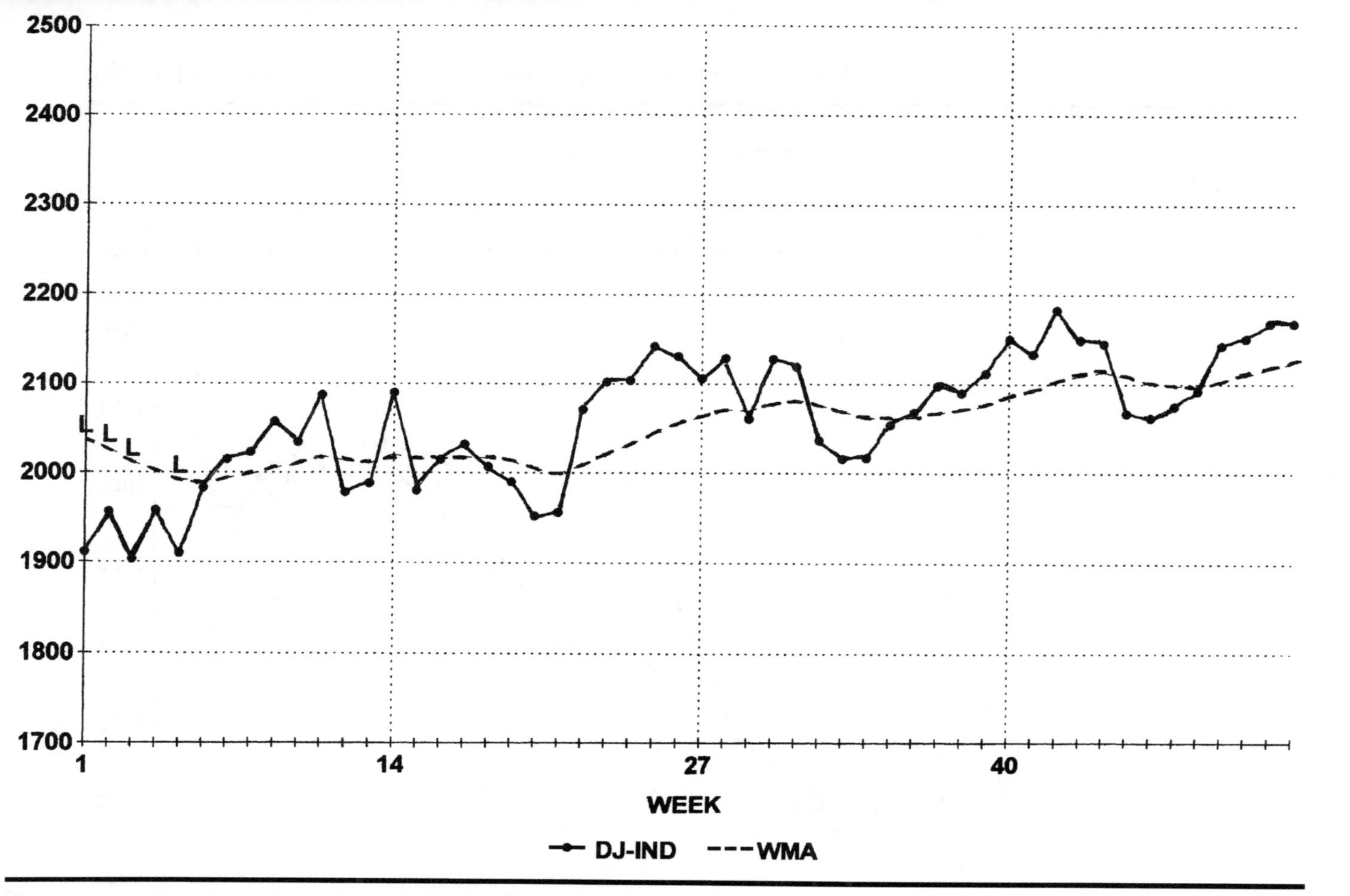

ure 8.3 DJ-Industrials 1988 GAP from Weighted Moving Average

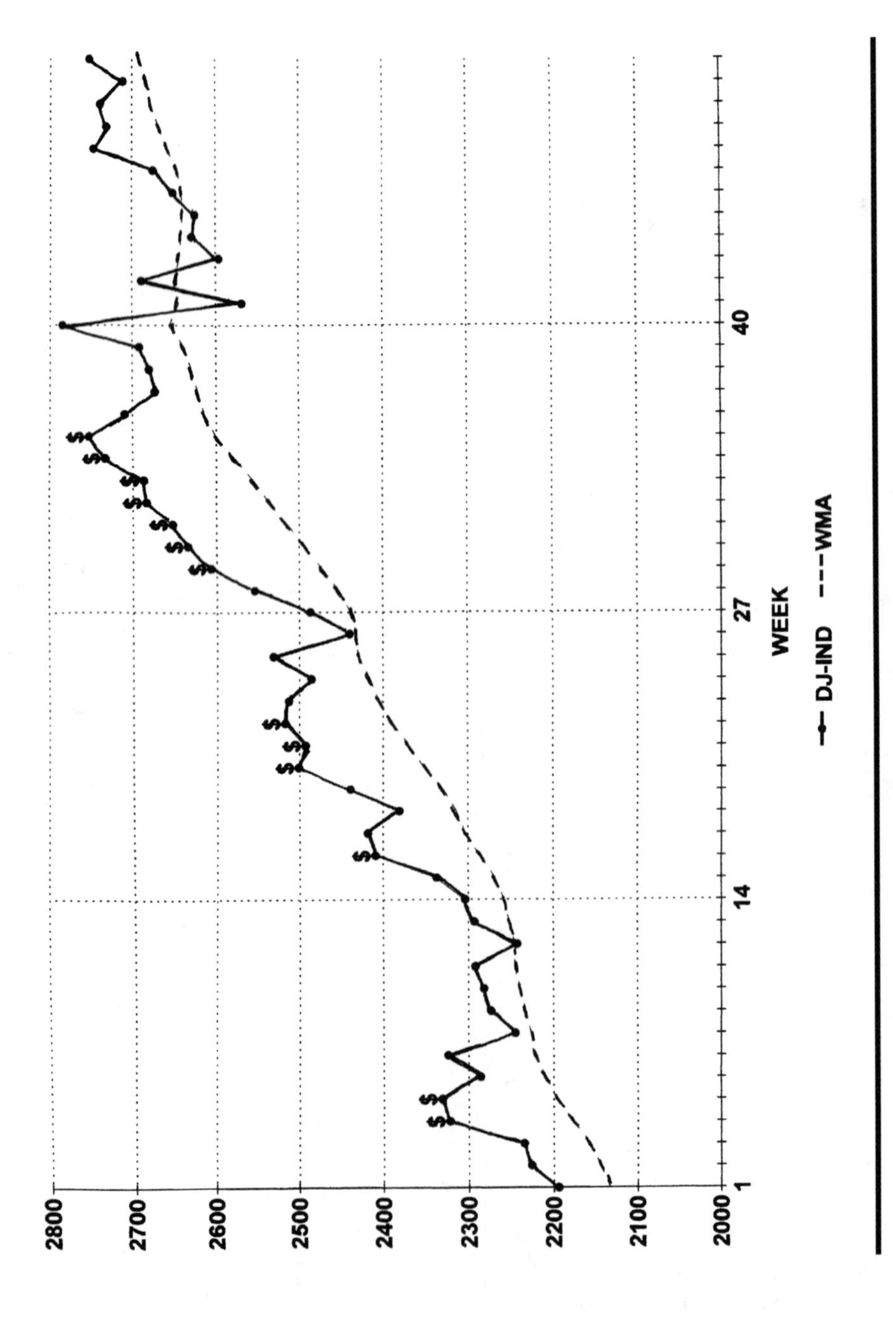

Figure 8.4 DJ-Industrials 1989 GAP from Weighted Moving Average

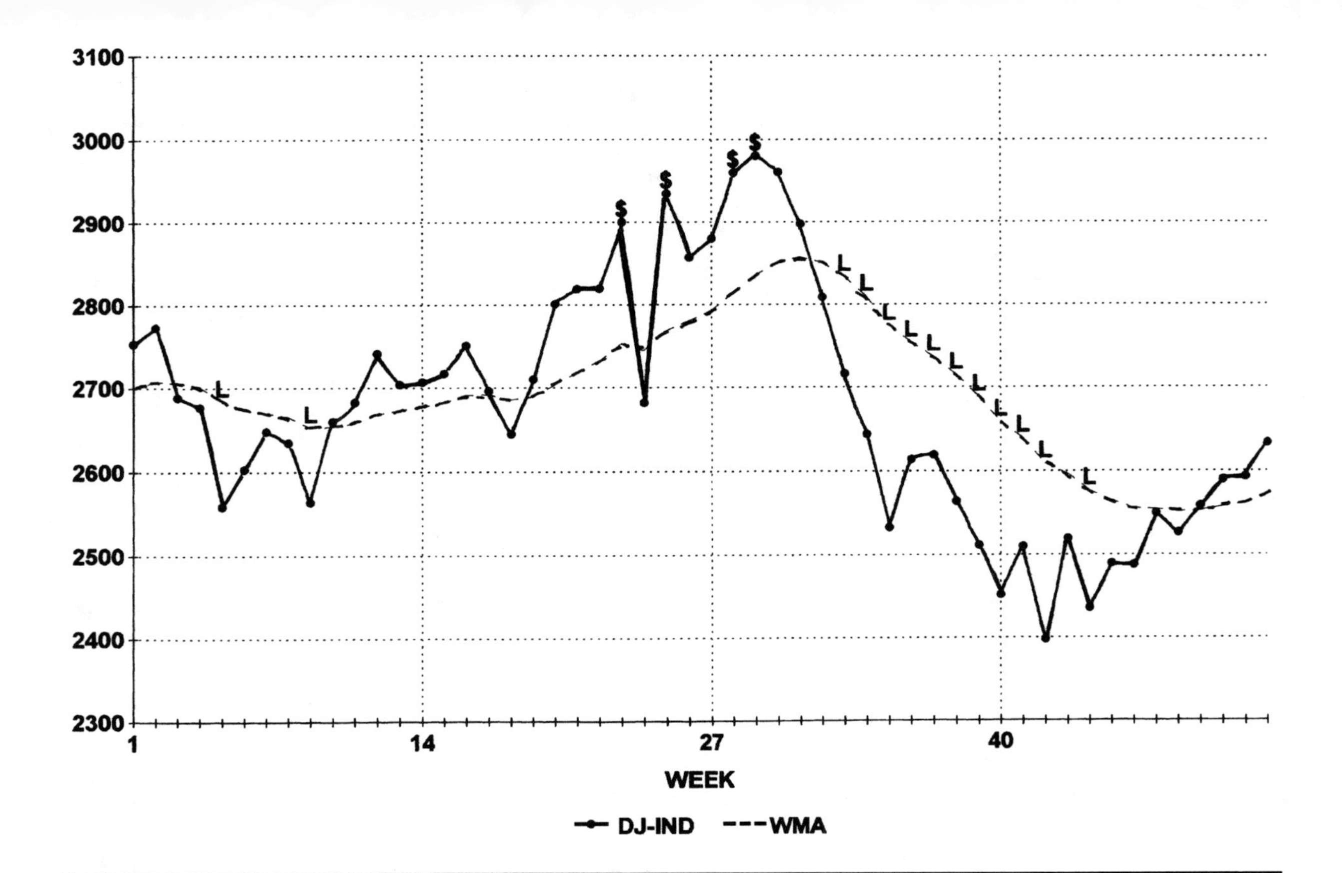

Figure 8.5 DJ-Industrials 1990 GAP from Weighted Moving Average

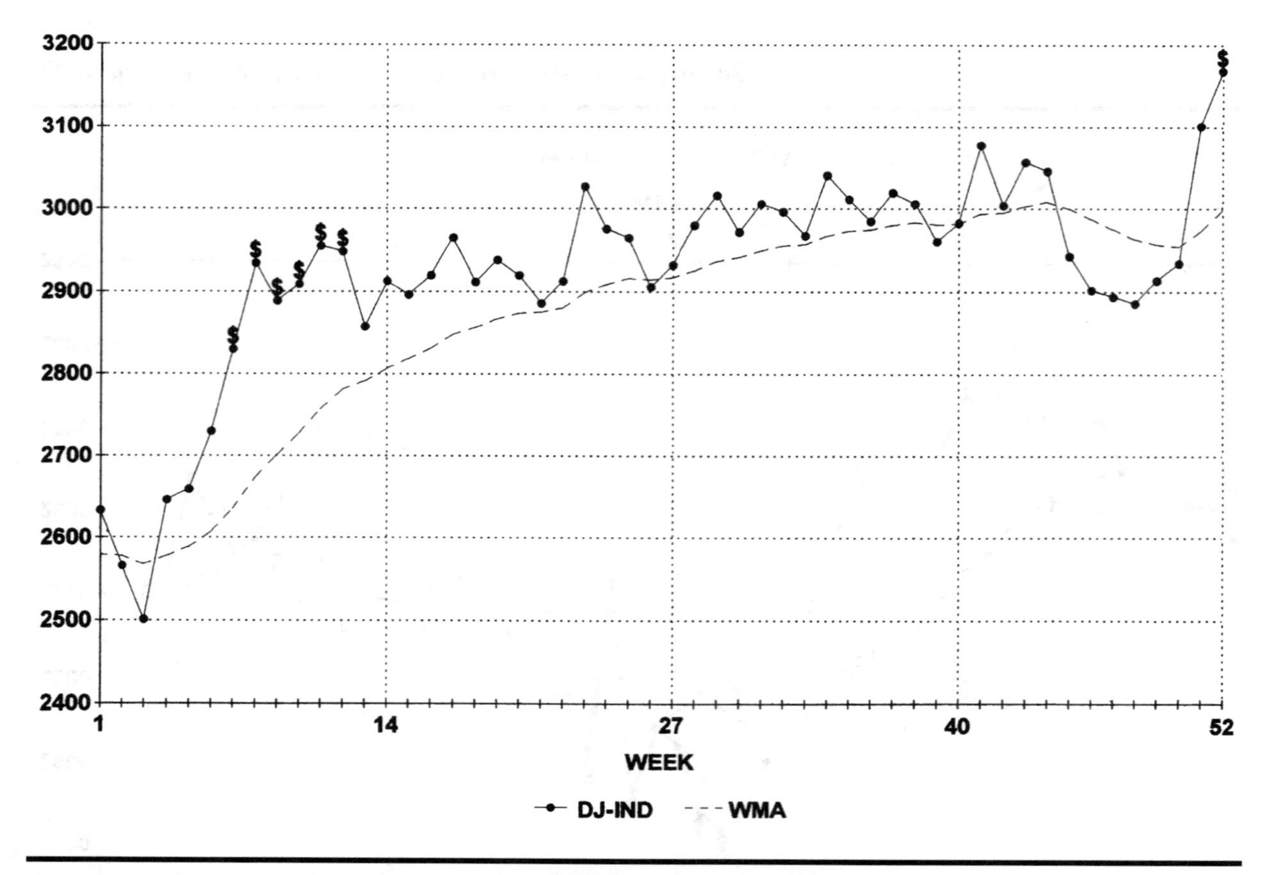

Figure 8.6 DJ-Industrials 1991 GAP from Weighted Moving Average

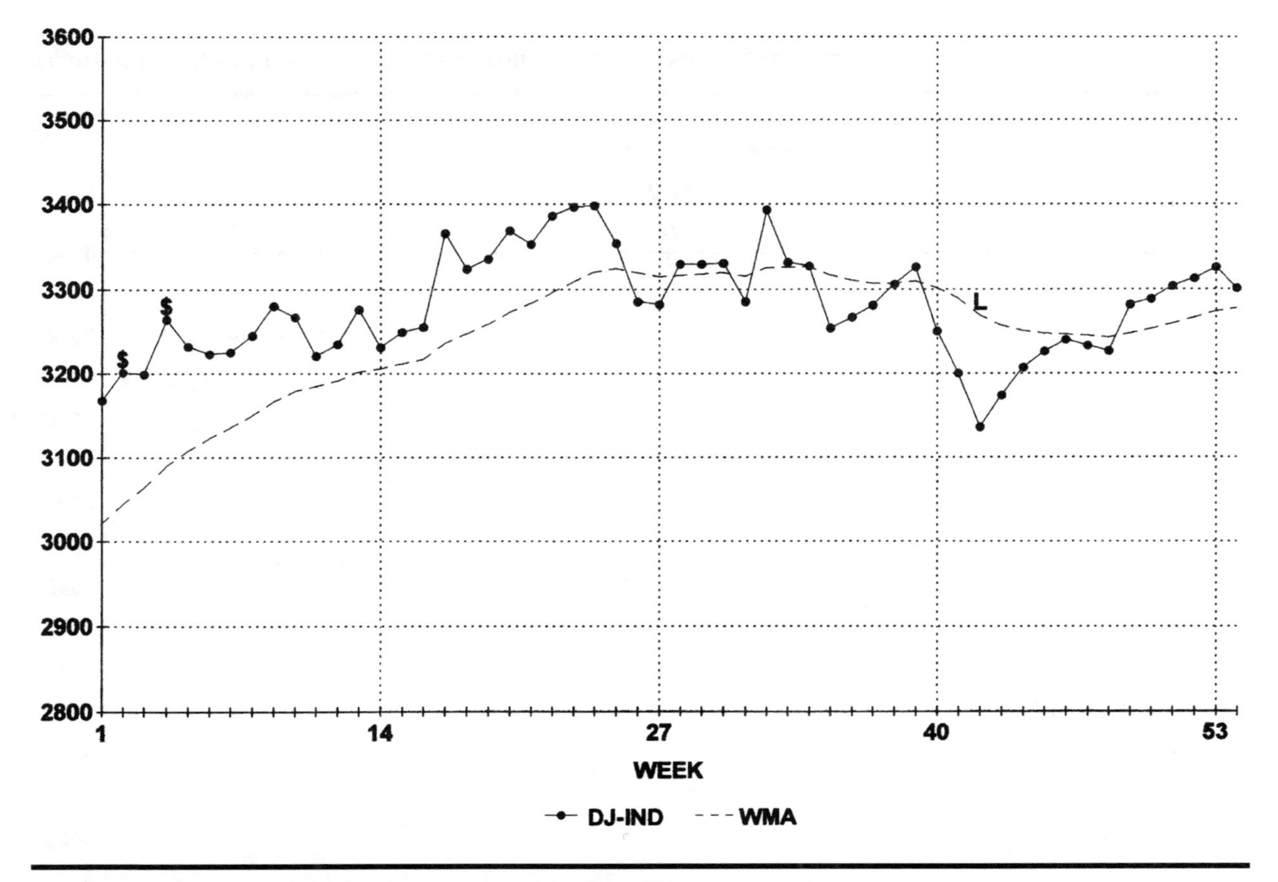

Figure 8.7 DJ-Industrials 1992 GAP from Weighted Moving Average

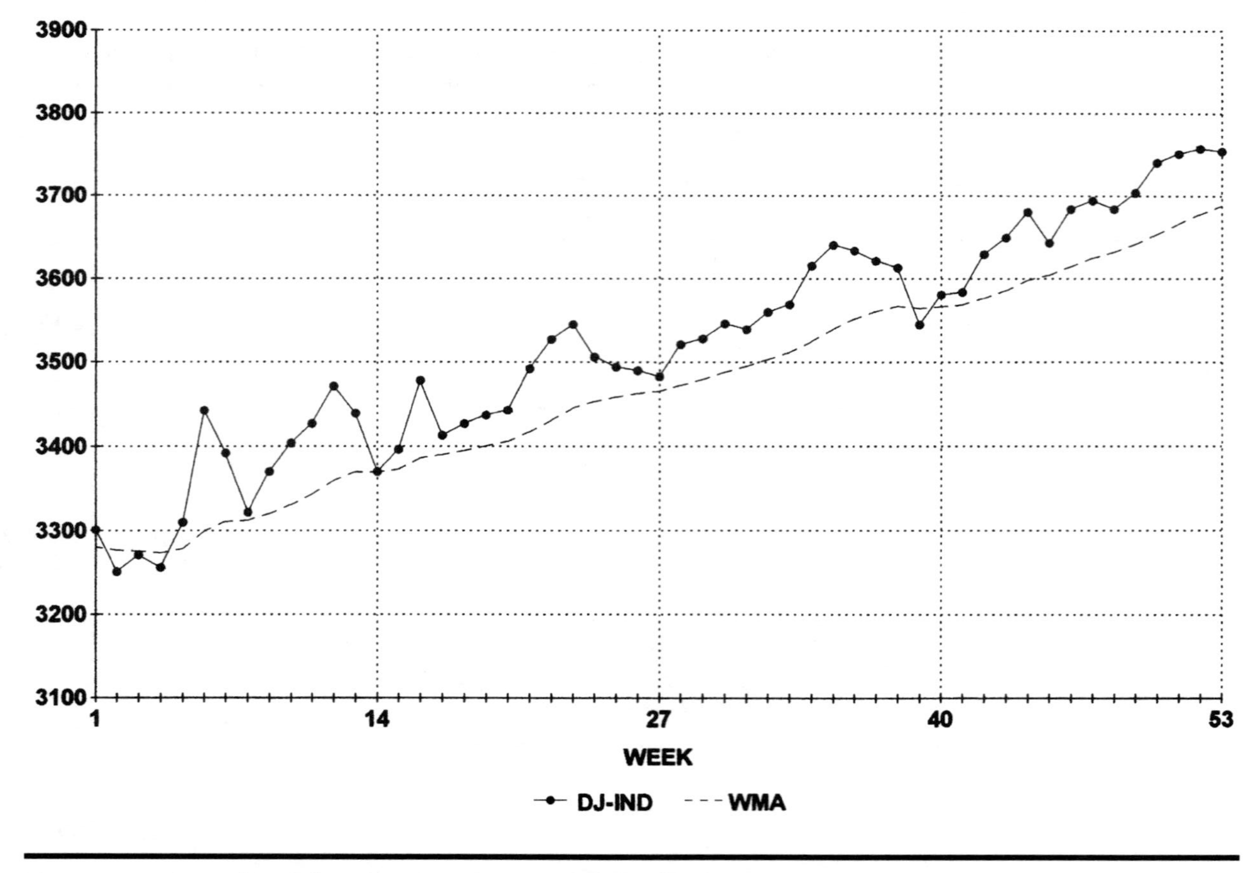

Figure 8.8 DJ-Industrials 1993 GAP from Weighted Moving Average

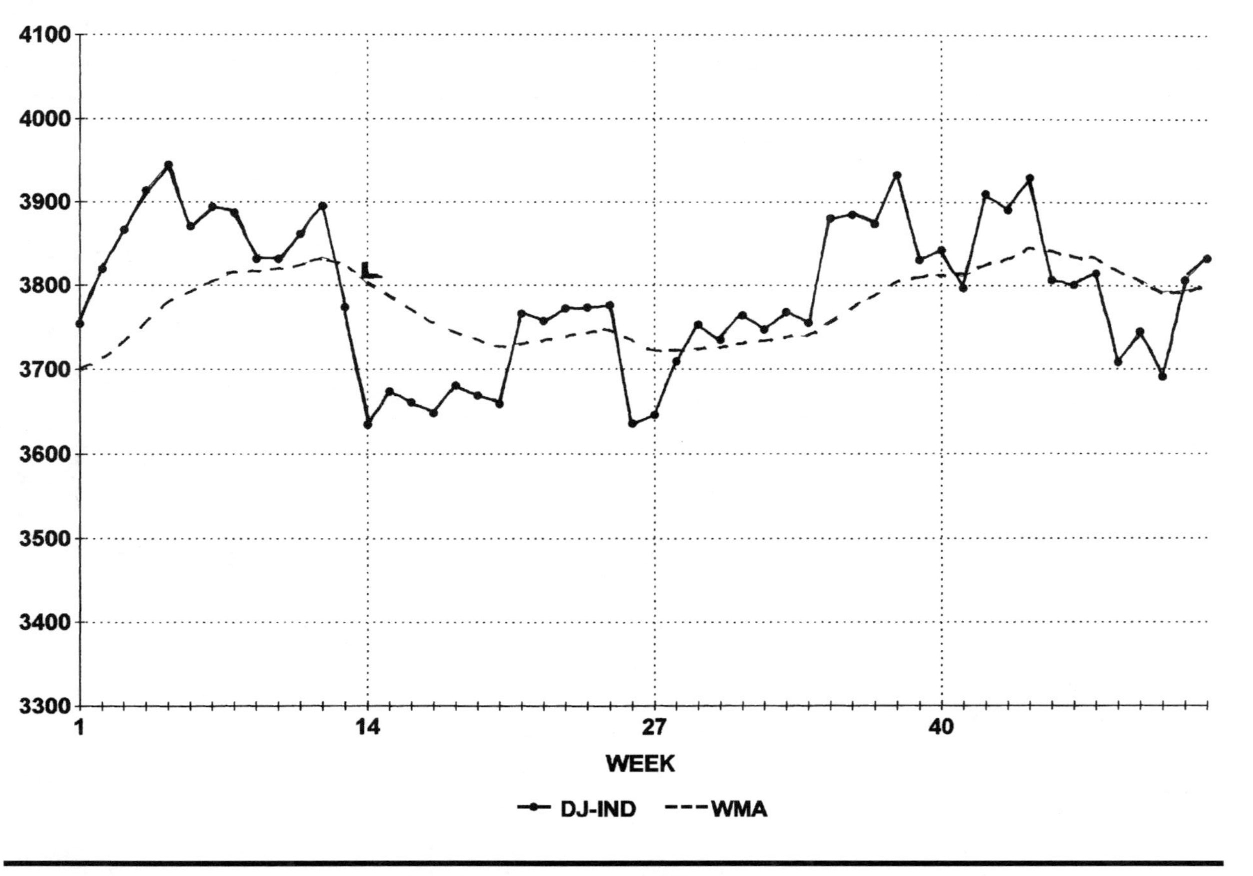

Figure 8.9 DJ-Industrials 1994 GAP from Weighted Moving Average

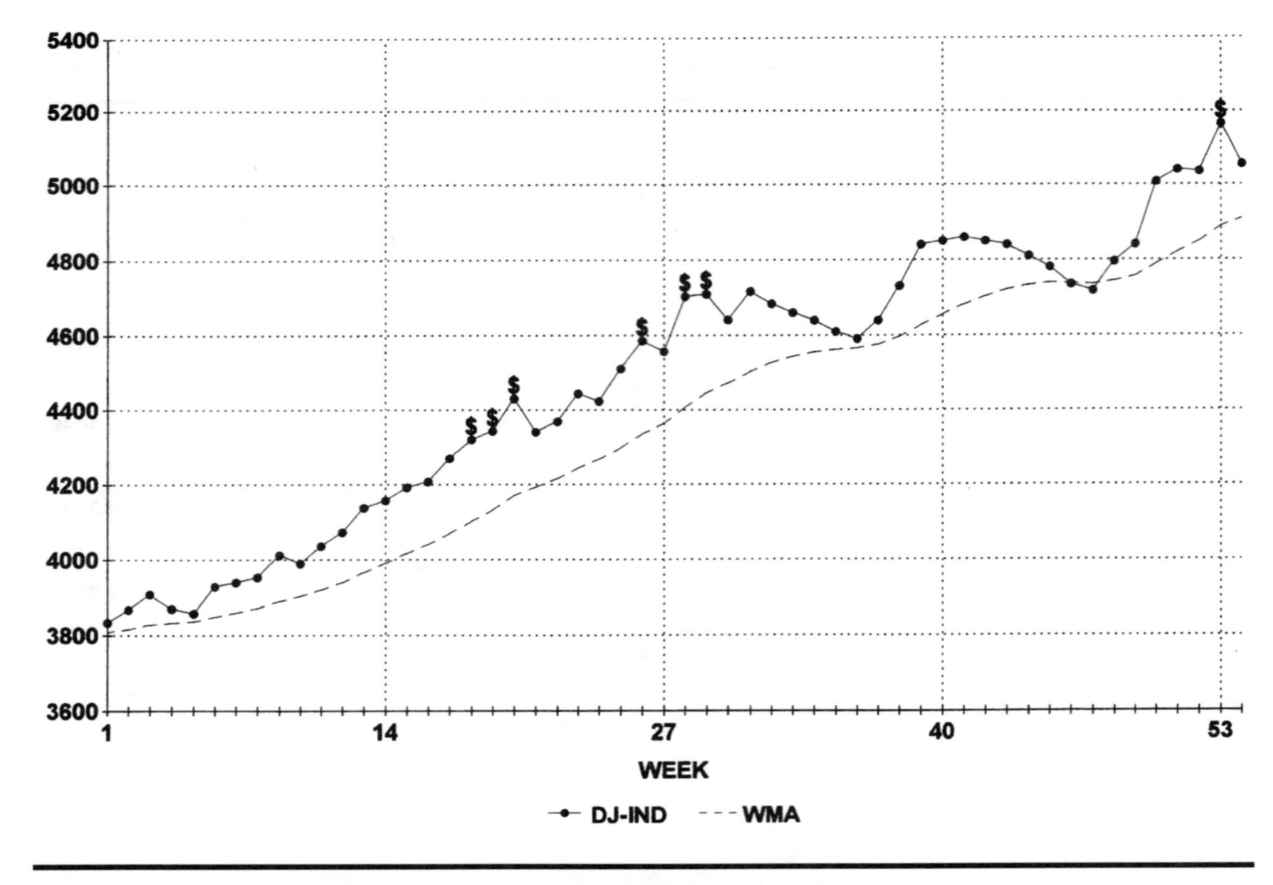

Figure 8.10 DJ-Industrials 1995 GAP from Weighted Moving Average

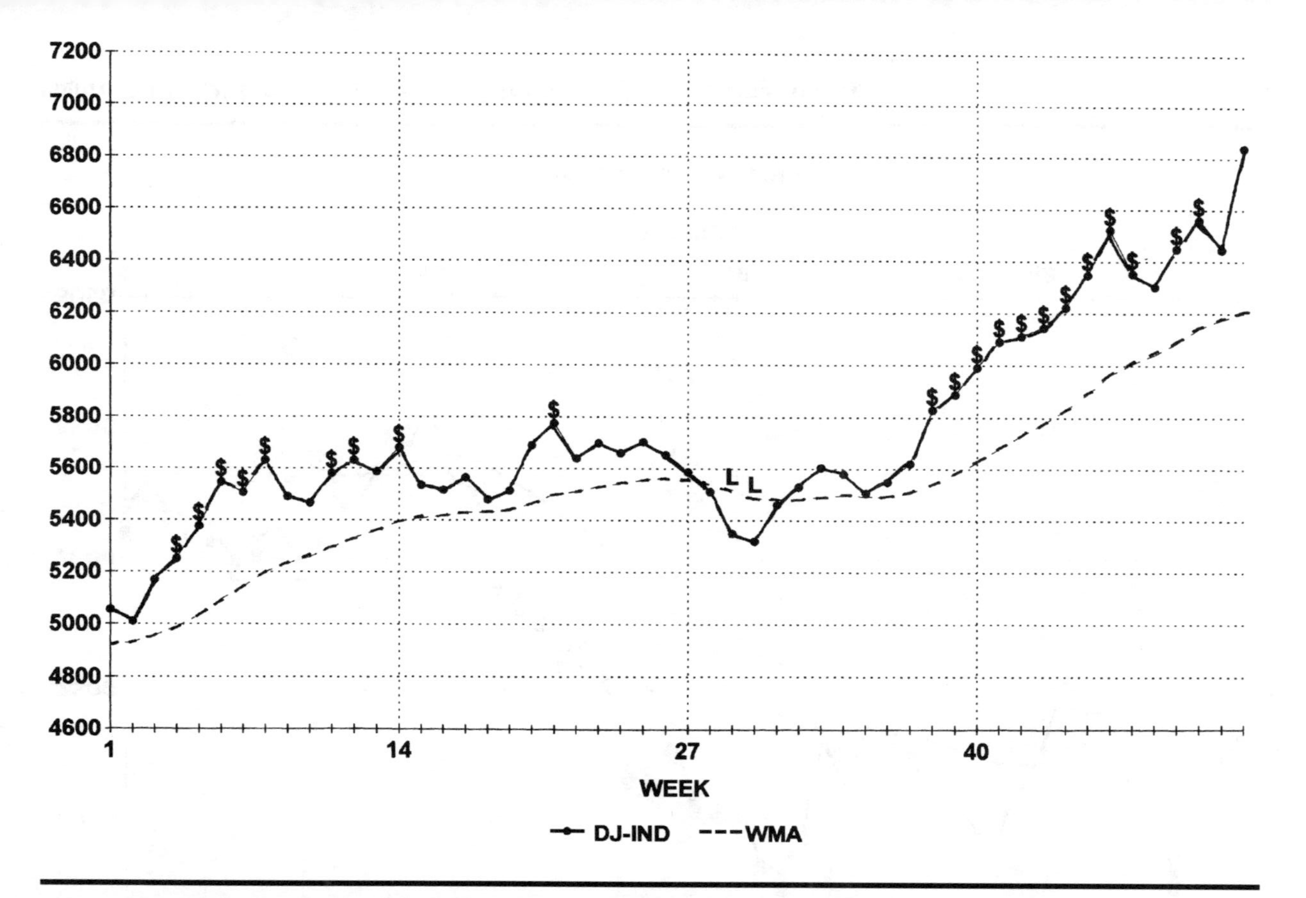

Figure 8.11 DJ-Industrials 1996 GAP from Weighted Moving Average

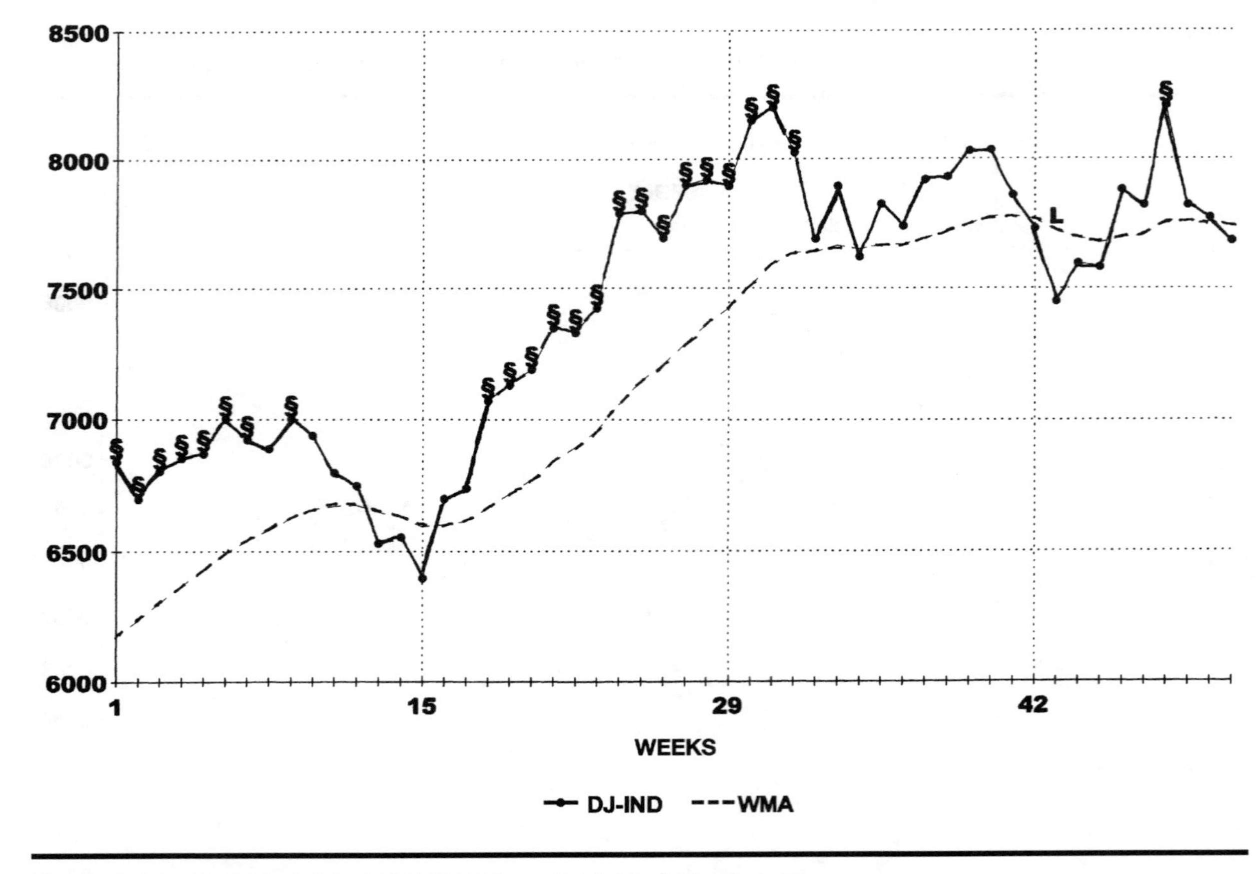

Figure 8.12 DJ-Industrials 1997 GAP from Weighted Moving Average

9 Bond Fund Investment Indicators

By investing in bond mutual funds, investors become lenders to governments, corporations, or municipalities that need money to finance their expansion programs or debt. As indicated earlier, any investor in bond funds must be aware of the type of bonds the fund is predominantly invested in. The average bond rating and average maturity of the bonds in the fund are key factors to consider.

9.1 Debt

Let's consider an example of personal debt and then a nation's debt. Consider the following:

A) Joe: Income 180,000 a year, debt of 108,000; secure job,
B) Henry: Income 30,000 a year, debt of 18,000; job not secure.

Clearly, Joe's debt is six times higher than that of Henry's and thus might be construed as Joe being in a riskier financial situation. This despite of the fact both have the same percentage debt compared to their income. But, now ask yourself the question: which of the two more likely will be able to pay back his debt without placing himself into too much of a constraint? Clearly, Joe will be in a better position to do so because he produces a six times higher income than Henry.

A study by a Swiss financial paper delivered some surprising results when this same line of evaluation was applied to the debt load of individual countries.

	Debt as % of GNP	*Debt per capita ($)*
U.S.	152	47,800
Germany	166	50,400
Japan	210	91,300
Great Britain	216	58,000
Switzerland	224	96,500

Data from *Cash*, Switzerland, June 1995.

In other words, the known "champion saver" nations Japan and Switzerland are worse off than the U.S. when measured both in Debt/GNP and Debt/Capita terms. Although this data is from a few years back, it is nevertheless of value since we can with certainty assume the debt loads have not decreased in the meantime.

There are essentially two groups of bond fund investors. The first sets its focus primarily on the steady stream of interest the funds pay on a monthly or quarterly basis. Many retired people belong to this group; they tend to remain invested in bond funds irrespective of gyrations in the fund's Net Asset Value (NAV).

There is however a large difference between investing in a bond fund and an individual bond itself, a factor quite often overlooked by many mutual fund investors. An investor in an individual bond can more or less rely on a steady stream of interest income *and* expect to receive the full face value of the bond at its expiration date. Not so the bond fund investor. Bond fund managers very seldom keep the bonds in the portfolio to full maturity; they frequently turn over the portfolio by selling existing and buying new bonds with the potential for better returns. A bond fund thus will experience changes in NAV similar to equity funds. However, in bear markets, bond funds tend to decline less than equity funds, the reason being that bond interest payments cushion the effect of the downturn. This flight from equities into government bonds can be most often observed when the equity markets severely correct in a short period of a few days. This switch into bonds is then often referred to as a "flight to quality." The question remains if the such viewed bonds can indeed be classified as "quality" in view of earlier discussions on this subject of government fiscal actions.

In the other group we find the investors that buy and sell bond funds in the same manner as equity funds. These are investors who generally concentrate on the total return performance and will sell when the performance of the bond funds is not meeting their goals and expectations. Since bond funds have their ups and downs like equity funds, and their performance most often is in lock-step with the equity markets, our investment strategy follows the line of thinking of the second group.

Prices of bond funds predominantly react to changes in interest and inflation rate trends. Hence, bond funds are one of the simplest types of investments to control. Quite simply stated, when interest rates are on the rise, it is not opportune to be invested in bond funds. The best total returns in bond funds can usually be obtained when interest rates are in a downward trend.

9.2 Bond Ratings

Of very significant importance is the rating of the bonds contained in the fund. A fund that contains predominantly high rated bonds is less risky, however also offers lower yields. High income bond funds (so-called junk bond funds) carry more risk because the fund is structured around lower rated bonds and thus must offer higher yields to attract investors. Junk bond funds have larger potentials for both up and down side price movement when compared to conventional income funds. In the middle of 1998, 10-year government bond yields were 5.4%, company bonds (AAA rated) paid 6.1%, and junk bonds (B rated) a whopping 9.75% interest. The yield of these three types of bonds are directly related to their risks. Since the government guarantees to pay the full principal back after 10 years, it can peddle its debt papers to investors at the lowest rates. There remains only the question what the principal is worth in buying power after the ten years. In this respect, Russia did send the investment world some nasty greetings in 1998. When the government didn't have the liquidity to repay their debt papers, they just turned up the printing presses in the hope inflation would take care of the financial crisis. The Russian ruble was significantly devalued; in short, they rolled the problem over to the investors and their own citizens. In contrast, in hindsight we find many companies that issued junk bonds, despite their low ratings, to have been less risky than some foreign (or domestic) government bonds.

An investor is well-advised when told not to chase bond funds with the highest yields. We should also not blindly assume our own government's bonds are risk free because we believe to live in the greatest and most powerful country in the world. Things can change rapidly. Bond funds are thus not buy and hold investments. They have to be carefully evaluated before we commit any of our money.

9.3 Indicators to Assess the Bond Market

The first indicator we consider for bond fund investments is the current bond interest rate trend. We use the 10-year Treasury Bond Yield as a yardstick for current interest rates. "Current trend" in our terms means the change in rates over the past month. Specifically, we want to know if the trend is up (= high in our Fuzzy Logic diagram), down (= low), or remains stable (= ok).

If interest rates moved up more than 0.4% = High
If interest rates moved less than 0.4% = OK
If interest rates moved down more than 0.4% = Low

The second indicator is the current inflation rate trend. The Consumer Price Index (CPI) is the data we use for this purpose of analysis. We look at the trend of this indicator over the past month.

If inflation rates moved up more than 0.4% = High
If inflation rates changed less than 0.4% = OK
If inflation rates moved down more than 0.4% = Low

The inflation rate indicator has already been used for the evaluation of the current economic condition and can thus be entered quickly in the Fuzzy Logic diagram for bond fund investments (Figure 9.1).

A trend must be confirmed before we consider it real. This is the reason why we have selected a 0.4% change as being significant. Moves by the Federal Reserve Bank are more often in steps of 25 basis points (0.25%), two consecutive moves of this magnitude are considered by us to be significant. Daily fluctuations or a single government report issued during a given month might not necessarily constitute a trend change. However, when a trend change according to our criteria is indicated, we will evaluate the merits of making changes in our portfolio's bond fund holdings. Again, as previously with equity fund investments, these portfolio changes must be in line with the personal maximum risk level established for the investment portfolio.

With two indicators considered, each being either HIGH-OK-OR LOW, our Fuzzy Logic concept will thus deliver a total of 3 × 3 = 9 basic bond fund investment conditions as shown in Figure 9.1 (where "T-Bond Rate" stands for Treasury Bond Rate).

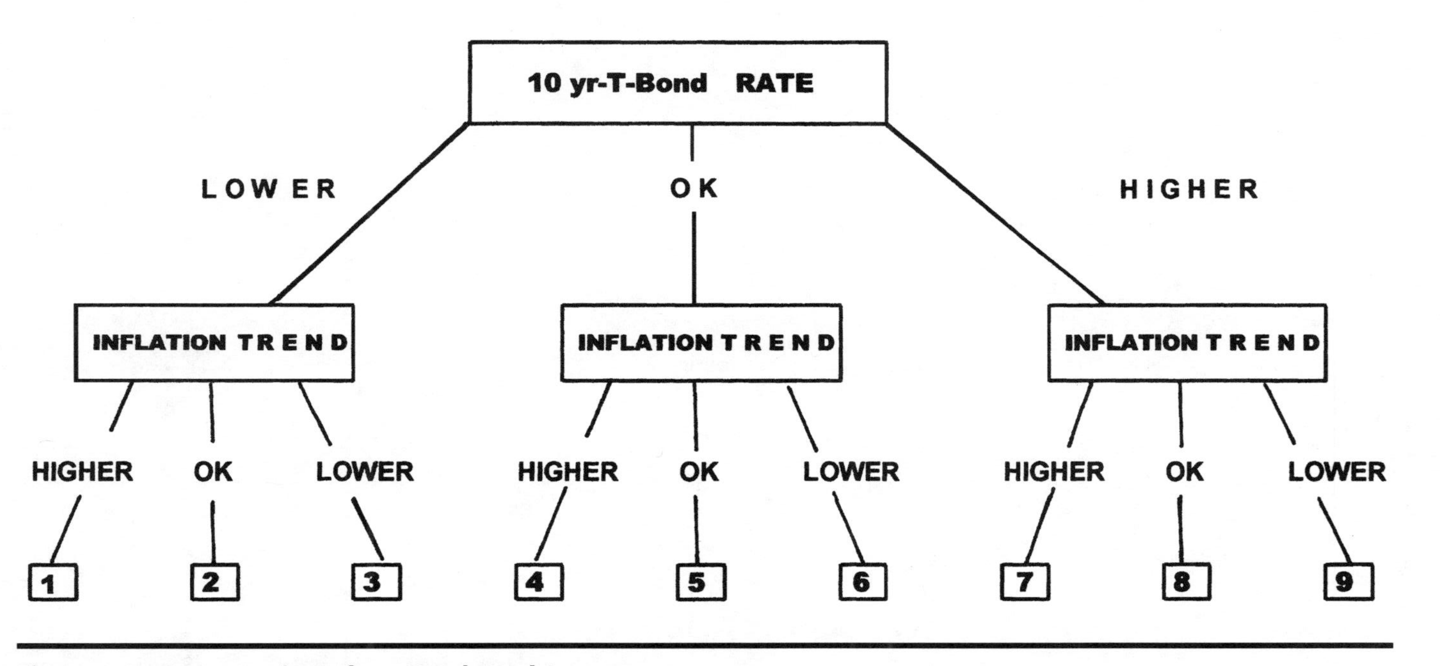

Figure 9.1 Fuzzy Logic Path — Bond Fund Investments

10 Indicators for Foreign Investments

Mention was made at the onset of this book that the investment markets of the Western industrialized nations are now closely interwoven with each other. The markets are now global in scope. London, Frankfurt, Tokyo, Zürich, and other markets are sensitive to what happens on Wall Street and vice-versa. We have all become dependent on each other for trade, the flow of currencies, and each other's economies. We cannot afford to ignore investment opportunities overseas and must therefore become familiar with the world outside our own borders. We must have an understanding of some of the fundamental factors that have brought about this globalization of the markets and the picture is not very flattering to the U.S. nor the Europeans. Whenever one talks about foreign investments or trade, one invariably has to address one's self to the subject of currency exchange rate volatility and particularly to what is happening to the value of the U.S. dollar.

10.1 International Trade

The U.S. has been on the forefront striving for free trade among all nations. To support its global goals, the U.S. has kept its doors wide open for foreign producers for the past 30 years. Unfortunately, these noble objectives have been clearly one-sided. America's trading partners have not been eager to follow that example for fear of losing their export advantage or the possibility of being flooded with less expensive American food products that would hurt their own farmers.

Nationalism is still very much alive. Like it or not, each government has its own priorities not the least of which is to take care of their own people first and foremost. Even if it means subsidizing whole industries and farm segments, charging excessive import duties to keep foreign products out of its borders, or maintain exorbitant prices for some of its farm products.

When U.S. trade delegations go to Europe to obtain concessions on agricultural issues, or when the chairman of a large U.S. car manufacturer raises Holy Cane with the Japanese, they usually get a polite reception and perhaps vague promises ending in foot-dragging, deliberate stall tactics, and stonewalling. By closing their own borders to foreign goods but being allowed to sell their own products almost unrestricted in the U.S., they created themselves something like a free ride. It will get worse before it will get better.

The European Community (EU) is actively imposing import barriers that borders on the ridiculous just to prevent goods from entering their territory. Everything imported to this trade block must meet EU specifications specifically designed to make it difficult for foreign producers to sell their products in Europe. There are specifications for bananas (length and degree of bend!!!), condoms, telephones, cars, electrical household appliances, specifications for just about anything Europeans produce at a much higher cost than some foreign producers. What kind of a trading partner is this?

10.2 The U.S. Dollar

On the other hand, the U.S. had a free ride for decades, too, when the U.S. dollar became the world's key currency. As a result, the world has been flooded with U.S. dollars. The enormous amounts of dollars floating around could sometime in the future haunt the U.S. Should people worldwide lose faith in the dollar and all at once dump their holdings, a major financial crisis of unprecedented proportions could erupt. Impossible? We don't think so. Just think what such a massive unloading frenzy would do to the value of the dollar. A massive devaluation could then not be avoided. The steady deteoriation of the U.S. dollar versus other currencies in the past 15 years would be nothing compared to this cataclysmic event. We don't predict; we also don't expect such events, but we must be aware that such possibilities do indeed exist.

There are now an estimated 500 billion U.S. dollars changing hands every day outside the U.S. borders. These dollars float from country to country and very seldom find their way back to the U.S. Other nations and corporations trade with each other and settle their balances in U.S. dollars, Eurodollars, Petro-dollars, and whatever they may think of next.

10.3 Currency Exchange Rates

This leads us directly to the crucial role the Currency Exchange Rate plays in matters of foreign investments. As the examples in Tables 10.1 and 10.2 show, a U.S. investor having purchased shares in a German fund could have netted a return of 11.6% despite of the German market losing 0.5%, but its

Table 10.1 Foreign Stock Markets and Currency Rates a) From an U.S. Investor's Viewpoint

	Stock Market	*Currency Exchange Rate*			*Net Change*
Country	*% Year-to-Date*	*Now*	*6 Months Ago*	*% Change*	*for a U.S. Investor*
U.S.	–3.5	1.0	1.0	na	–3.5
Japan	–13.5	135.6	133.7	–1.4	–14.9
Switzerland	–9.6	1.36	1.53	12.5	2.9
Britain	–4.4	0.5865	0.5982	2.0	–2.4
Germany	–0.5	1.65	1.85	12.1	11.6
Singapore	–40.1	1.69	1.62	–4.1	–44.2
Mexico	–35.2	10.39	8.5	–18.2	–53.4
Canada	–18.8	1.55	1.42	–8.4	–27.2
France	1.3	5.52	6.21	12.5	13.8

Table 10.2 Foreign Stock Markets and Currency Rates b) From a Canadian Investor's Viewpoint

	Stock Market	*Currency Exchange Rate*			*Net Change for Canadian*
Country	*% Year-to-Date*	*Now*	*6 Months Ago*	*% Change*	*Investor*
U.S.	–3.5	0.645	0.704	9.1	5.6
Japan	–13.5	87.484	94.155	7.6	–5.9
Switzerland	–9.6	0.877	1.077	22.8	13.2
Britain	–4.4	0.378	0.421	11.4	7.0
Germany	–0.5	1.065	1.303	22.3	21.8
Singapore	–40.1	1.090	1.141	4.7	–35.4
Mexico	–35.2	6.703	5.986	–10.7	–45.9
Canada	–18.8	1.000	1.000	na	–18.8
France	1.3	3.561	4.373	22.8	24.1

currency increased 12.1% against the U.S. dollar. In the other hand, a Canadian investor having placed some money into the Swiss Stock market, realized a gain of 13.2% although the Swiss market fell 9.6% but the Swiss franc increased 22.8% against the Canadian dollar. Tables 10.1 and 10.2 reflect currency exchange rates that prevailed in September 1998.

Clearly, currency exchange rate trends must be of prime consideration to any investor contemplating placing his money in a foreign equity or a bond fund. The best opportunity for foreign investments arises when both that foreign country's currency as well as its equity or bond market is on the rise. We must also keep in mind the down side risk of such investments; it can be just as dramatic.

In our investment strategy we attempt to take advantage of this so-called dual return potential in foreign investments. We don't want to be in a foreign fund when the currency and the market of the given country is on a decline. Mutual funds specializing in foreign investments carry an extra risk that is not found with domestic bond or equity funds. Since currency fluctuations play an important part in our strategy, we must become familiar with the forces that can drive the value of a currency in either direction.

Who makes the value of the U.S. dollar change? There have been theories forwarded accusing an elite group of bankers of manipulating the currency markets to their advantage. Another theory points the finger at greedy speculators controlling the world's currency options and futures pits. There is no question that a select few with enough power and financial clout, including some Central Banks, do trade heavily in currencies and that their actions have a certain impact on the markets. But, there are just as many speculators doing the same thing in the equity, bond, real estate, or any other markets. The free enterprise system allows such activities. Where there is a profit to be made, there will be private individuals, bankers, and corporations who want to take advantage of such opportunities. It is wrong to accuse these people of causing unrealistic currency exchange rate fluctuations. Often, governments impose currency trading restrictions in an attempt to prevent a free fall of their currency. Or Central Banks, individually or as a group, intervene in the currency markets to shore up a given currency. These actions, however, are only temporary solutions and usually not very effective for the long haul.

The fundamental reasons for losses of currency values are much deeper rooted. Governments with their ill-timed nationalistic priorities and fiscal policies are the prime causes for these changes. More specific: monetary policies carried out by Central Bankers under pressure from politicians, are the prime driving forces behind a currency's rise or fall. Governments know full well they must offer foreign investors higher interest rates or bond yields if they want to attract foreign money. A heavily indebted country, dependent on foreigners to finance their deficit spending will, under normal economic conditions, not lower their interest rates for fear of losing this capital inflow. This has happened in Germany after the reunification. Deliberately holding interest rates higher than the other industrialized nations, they were able to finance the enormous costs of this East–West German unification in part with foreign capital.

The United States has done similar things in the late 1980s and early 1990s. By offering Japanese and European investors higher bond yields than they could find in their own country, these foreign investors bought U.S. Treasury bonds and thus helped to finance the large U.S. budget deficit. We would not be surprised to see the same thing happening again when Europe embarks on the common currency, the Euro, in 1999.

In short, higher interest rates attract foreign capital and thus cause the currency to rise. Lower interest rates might compel foreign investors to withdraw their money and seek higher yields somewhere else in the world. Result: the currency tends to fall in value.

If foreign investors buy large volumes of U.S. equities or bonds, they need U.S. dollars to pay for these securities, causing the value of the dollar to rise. Conversely, if vast numbers of foreign investors liquidate their U.S. holdings, the value of the dollar will fall. Clearly then, concentrated actions by foreign investors also have an influence on the value of a currency.

America has been running a trade deficit for more than ten years. In other words, it imports more goods than it exports. Thus, more U.S. dollars are going overseas instead of circulating around within their own borders. Net result: increased pressure exerted on the value of the dollar. For every dollar that leaves the country, there will be a dollar less available to buy American goods and services.

We have learned earlier, governments will function only as long they have money to spend. With no money to spend, and unwilling to raise taxes, there remains only one choice: finance government expenditures by creating new debt. Borrow more money, preferably from foreign investors. Spend now — pay later. Even if it means raising interest rates to appease these foreign investors. In short, the problems are being swept under the rug by inflating the money supply.

Currency exchange rates periodically reach over- or undervalued levels. A situation can arise when goods and services in one country become much cheaper and thus, everybody wants to buy that country's products. With heavy demand from overseas for their goods, the country will prosper and its economy expands. After a while, workers there will become unhappy and start to demand higher wages. Also, with so much money inflow from foreign lands, people will have more money to spend. End result: inflation will start to waggle its nasty tail in that country. After a few years of rising wages and inflation, products from that country suddenly are not cheap anymore. The tide turns, and foreigners will buy less from that country. When this situation develops, the government springs into action.

With faltering exports and the possibility of a slowdown in the economy, it lets the value of its currency slide in order to make prices again competitive. Governments do this by selling their own currency in the open market. Lately, these currency adjustments are being made by consensus of the group of seven industrialized nations. They set target ranges within which a currency is allowed to fluctuate, intervening when any one steps out of this range. Intervention in the currency markets has proven to be a temporary excercise in futility.

In the long run, market forces will always prevail. If a currency is drastically overvalued, markets themselves, not governments, will ultimately restore the value to its rightful level.

10.4 Purchase Power Parity

Many experts in the foreign investment arena adhere to the purchase power parity (PPP) theory to identify over- or undervalued currencies. This theory is based on the assumption that a currency's value is related to the inflation rate differential between the two countries in question. If inflation is much higher in one country than the other, the former currency value will have to increase in relation to the other. This theory has much merit if applied to prices for import–export goods. However, it also has its flaws when not viewed in the proper context.

Take for example the cost difference in 1998 of some widely consumed products between the U.S. and Switzerland

	All Stated in Terms of U.S. $ Exchange Rate 1 U.S. $ = 1.40 Sfr	
	U.S.	*Switzerland*
1 cup of coffee	0.90	2.28
1 Big Mac & French fries	4.20	6.75
1 domestic soft drink	0.50	2.20

Based on the PPP theory one could possibly conclude that the Swiss franc is way overvalued and inflation much higher in Switzerland than the U.S. In matters of inflation, this is clearly not the case because Switzerland's inflation is 0.5%. Also, a Swiss can quite well afford these apparent high prices because his taxes are lower and his wages about double the wages in the U.S. Switzerland is expensive for visitors, but it doesn't export hotel rooms, Big Macs, and cups of coffee. The goods it exports, because of high wages, must be of very high quality to be able to command the necessary premium prices in the world's market places. A mutual dependency exists between quality and price of goods sold in any market. A deficiency in either could make the product noncompetitive. One cannot produce an inferior product and sell it at higher prices than what the foreign consumers are willing to pay. The name "Made in Switzerland, Japan, Germany, or U.S." doesn't carry the same respect anymore as it did twenty or thirty years ago. When foreign consumers notice the quality of given products is not what it used to be but prices are still inordinately high, they will turn their purchasing clout toward other sources where the price/quality ratio is still in order.

Purchase Power Parity (PPP) however has nothing to do with the wages earned and the cost of living in specific cities around the world. To demonstrate how the currency exchange rate can influence cost situations from a foreign viewpoint in a given country, we used the UBS's (Union Bank of Switzerland) periodic survey of living costs in over 50 different cities around the world. Tables 10.3 and 10.4 show excerpts of these data, published in 1994 and 1997, for some selected cities. All data shown are expressed in terms of U.S. dollars at 1994/1997 exchange rates. It is not difficult for the reader to determine how dramatically these figures have changed in the relatively short period of 3 years. What if the dollar would lose say 30% of its value over the next 10 years? Costs in the cities others than Houston would increase

Table 10.3 Wages and Cost of Living in Selected Cities — 1994 (all expressed in U.S. dollars at 1994 exchange rates)

	Zürich	*London*	*Dubli n*	*Housto n*	*Tokyo*	*Rio*	*Mex.Cit y*	*Toronto*	*Sydne y*
Annual Working Hours	*1874*	*1880*	*1727*	*1964*	*1893*	*185 7*	*2094*	*1888*	*1847*
Monthly Costs									
Food Basket	570	291	313	362	938	244	253	300	262
Services	430	300	270	280	460	260	220	220	230
Clothing (men medium)	580	455	400	605	1855	305	490	470	745
3-room Apartment Rent	950	1120	320	430	1730	880	730	690	510
Hourly Wages net	17.3	7.1	7.1	11.2	15.3	2.1	2.6	9.8	7.2
Monthly Wages net	2702	1112	1022	1833	2414	325	454	1542	1108

Table 10.4 Wages and Cost of Living in Selected Cities — 1997 (all expressed in U.S. dollars at 1997 exchange rates)

	Zürich	*London*	*Dublin*	*Houston*	*Tokyo*	*Rio*	*Mex.City*	*Toronto*	*Sydney*
Annual Working Hours	*1876*	*1839*	*1782*	*1875*	*1799*	*1892*	*2302*	*1927*	*1777*
Monthly Costs									
Food Basket	551	364	349	394	754	325	244	288	349
Services	430	350	320	290	450	350	220	260	290
Clothing (men medium)	720	670	620	900	1250	420	580	780	920
3-room Apartment Rent	930	1870	800	450	1360	980	730	540	590
Hourly Wage net	17.30	8.90	8.50	13.00	15.50	4.20	1.60	10.00	8.90
Monthly Wages net	2705	1364	1262	2031	2324	662	307	1606	1318

From: Union Bank of Switzerland, *Prices and Earnings around the Globe,* 1997 Edition.

by 30% from an American viewpoint while costs in Houston would be 30% less for a foreign tourist visiting Houston. Naturally, there are certain pitfalls in making such comparisons as is clearly stated in the UBS report. For example, food costs are based on European nutritional habits while the food basket in Mexico City would be clearly different. Quality of life is also not considered in that survey. Since the other scenario could just as well take place, namely, the dollar increasing 30% in value against all the other currencies, Houston could become an expensive city for visitors, but Americans with a stronger dollar, could once again face reasonable prices overseas. It doesn't do much good to complain how much more expensive goods are somewhere else because the reason for this is not greedy foreigners taking advantage of tourists, but we say it again, misguided fiscal policies of one's own government.

These are clear examples to show how important exchange rates are when considering investments in foreign countries.

10.5 Indicators for Foreign Market Conditions

To evaluate the opportunities for foreign investments, we look at two indicators: first, the bond or equity market trend in the country or region we are interested in to invest. The second indicator is the currency trend of that country or region when compared to the investor's residence country. For example, an American investor planning to invest money in a Japanese equity fund would a) determine the current trend of the Nikkei averages and b) evaluate the trend of the Yen compared to the U.S. dollar. Likewise, a Swiss interested in the U.S. market would look at the Dow Jones Industrials trend and check the trend of the U.S. dollar against the Swiss franc.

We use a worksheet as shown in Table 10.5 to keep track of foreign market and currency trends. The data needed for this analysis can be found daily in most financial publications. One must be careful to use the correct type of exchange rate reported since most publications will show two types of rates. It's not complicated. For example we might read: 1 U.S. $ = 1.36 Swiss francs and 1 Sfr = 0.735 U.S. $. Note: the second data is just the inverse of the first (1/1.36= 0.735).

Let's assume two scenarios where each investor purchased funds in the value of 21,609 U.S.$ each

Table 10.5 Foreign Markets and Currency Trends

	Month:	*Previous Month*	*Current*	*Market Trend*	*Currency Trend*	*Net Trend*
Japan	Nikkei					
	Yen					
Singapore	Straits					
	Times					
	S $					
Hong Kong	Hang Seng					
	HK $					
Germany	Dax					
	DM					
United Kingdom	FTSE-100					
	Pound					
Switzerland	SMI					
	Sfr					

Note: Enter foreign currencies in terms of 1 U.S. $. Market Trend = (current – previous)/previous * 100. Currency Trend = (previous – current)/previous * 100. Net Trend = Market Trend + Currency Trend.

1. 3 months ago a U.S. investor purchased 50 shares of a Swiss fund at a cost of 588 Sfr per share. The exchange rate at that time was 0.735 U.S.$/Swiss franc. Thus, his costs were: (50 * 588 * 0.735) = 21,609 U.S.$.
2. At the same time, 3 months ago, a Swiss investor purchased 250 shares of a U.S. fund at a cost of 86.436 U.S.$ per share. (Costs in terms of U.S.$: 21,609$). The exchange rate was the same, but the inverse of the above, namely, 1.36 Sfr/U.S.$. His costs therefore (250 * 86.436 * 1.36) = 29388 Sfr.

Both investors look in the papers now and notice the NAV of the shares has not changed for either fund but find the new exchange rates today to be: 1 U.S.$ = 1.538 Sfr and 1 Sfr = 0.65 U.S.$.

Conclusion: because of the shift in exchange rate, the U.S. investor's holding in the Swiss fund has lost 11.6% and is now worth (50 * 588 * 0.65) = 19,110 U.S.$. The Swiss investor's holding in the U.S. fund has gained 13.1% and is now worth (250 * 86.436 * 1.538) = 33,234 Sfr. All this just because the U.S.$ has gone up in value against the Swiss currency.

The author has evaluated and contemplated the use of other indicators for the purpose of foreign investments. One of these was the balance of trade surplus or deficit because there is strong evidence this figure influences currency exchange rates. Another indicator considered was the debt-to-GNP ratio both for the domestic and the foreign country. There is no question that changes in these indicators at some time or another can influence the currency exchange rate. But, the author has concluded the current currency trend seems to be the most logical one to use for our purposes. Again, in our Fuzzy Logic concept, simplicity is the virtue because the more indicators one uses, the more cloudy the overall important picture would become.

10.6 Tracking Foreign Currency Exchange Rates

The currency trend is a very powerful indicator for our purposes because unlike the equity markets, currency values don't change dramatically from one week to another. Currency exchange rates follow a certain trend over a relatively long period of time that can extend to months, quarters, or even years thus giving the foreign investor enough time to watch his investments overseas without having to make hasty buy or sell decisions when the currency trend reverses itself. One might argue that currency exchange rates tend to neutralize in the long run and thus have little effect on total return if a fund is held for five or more years. Statistical analysis of long term total fund returns might prove this to be correct, but it does not apply to our investment technique. Our horizon that decides if we buy, hold, or sell foreign funds is much shorter and we react when a trend reversal in currency exchange rates takes place.

We cannot apply the same yardsticks for foreign equity fund investments as we do for domestic equity funds. For example, P/E ratios in emerging markets have a completely different meaning. These ratios are usually much higher than in the industrialized nation's markets. The same applies to inflation rates. Thus, it would be wrong to stay away from investing in these markets just because the P/E ratio or the inflation rates overseas are higher than in our own market. A good investment opportunity in emerging growth markets could be lost. On the other hand, these emerging markets can be very volatile with extremely large swings to the up as well as the down side. A classic example were Southeast Asian equities, the darlings of foreign investors in the early 1990s. Losses of 50 to 70% in the aftermath of those booming years were not uncommon once the bubble burst. Because of the inherent volatility and being conservative investors, we usually limit our exposure to foreign investments to not more than 10% of total portfolio assets.

Thus, we again have $3 \times 3 = 9$ different basic conditions in our Fuzzy Logic diagram for Foreign Fund Investments as shown in Figure 10.1.

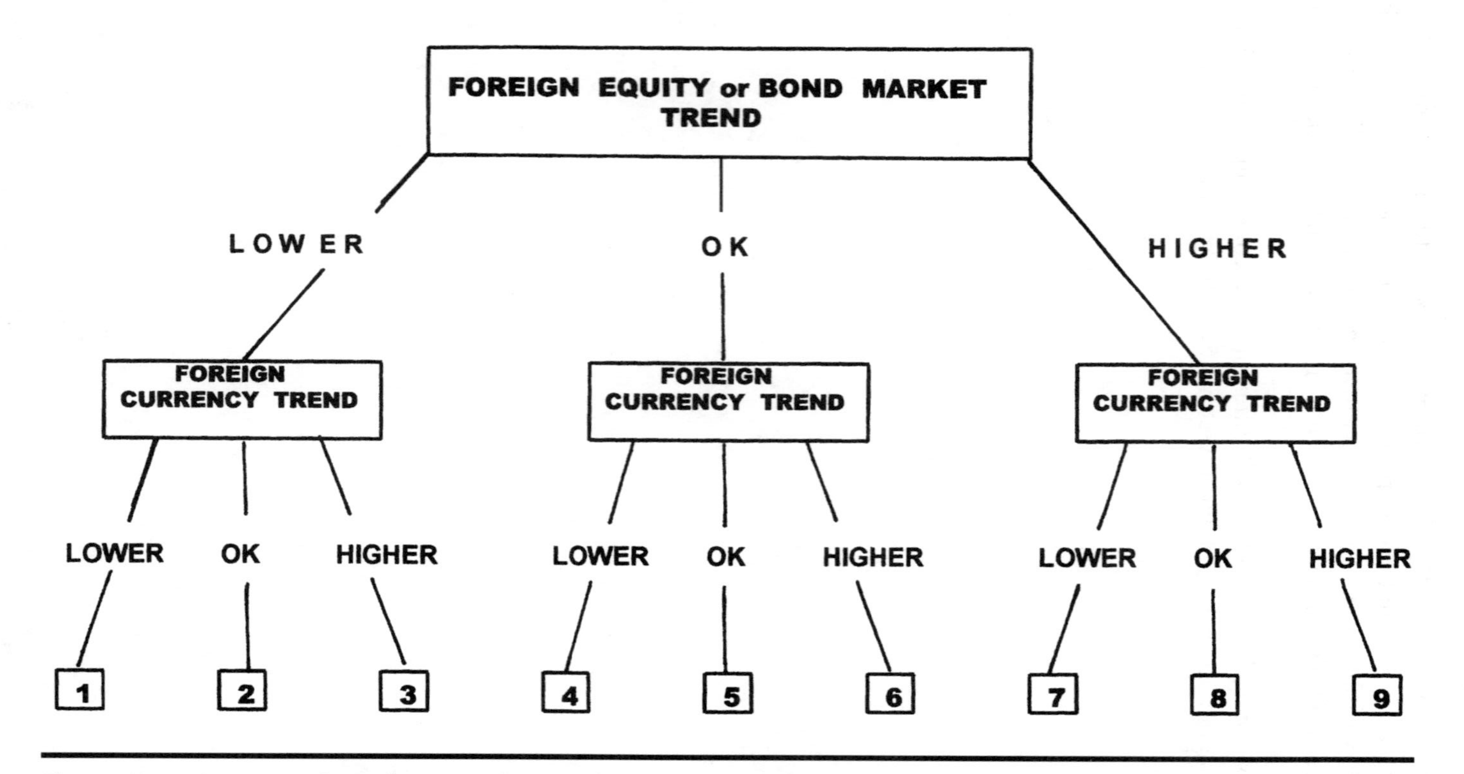

Figure 10.1 Fuzzy Logic Path — Foreign Fund Investments

11 Indicators for Gold Fund Investments

11.1 Gold's Value Through History

Throughout history, wars have been fought, nations conquered, and fortunes made or lost, all for the sake of holding the glistening precious metal, gold. A nation's wealth used to be measured in terms of how much gold it held in its treasury. Birthday gifts for kings were made of gold, who were even buried in golden caskets. Gods of distant and present cultures were permanently enshrined in the form of golden statues and still today people ornament themselves with all varieties of gold jewelry because gold is durable and has always been a symbol of wealth. International trade in ancient times was settled in gold bullion before governments started to issue paper money.

Then, sometime in the distant past, governments decided it was much more convenient to settle trade transactions in the form of paper money. In 1944, the U.S. dollar became the world's leading paper currency thanks in no small part to its military and industrial dominance during and after World War II. The Bretton Woods Agreement was signed, fixing all the world's currency exchange rates in terms of the U.S. dollar. In turn, the dollar was pegged to gold at a fixed rate of 35 U.S.$ per ounce. At any time, foreign governments could exchange their paper dollar holdings at this fixed exchange rate. In other words, the dollar, being on this fixed gold standard, had become *the* hard currency in the world. Most other world paper currencies were then also backed up by a given percentage of gold reserves.

The Vietnam war changed this international arrangement. President Johnson persuaded Congress to remove the gold backing clause to allow the Unites States to finance its war expenditures with newly printed paper money not pegged to gold. For a while, a two-tier market was created where the gold standard for government and Central Bank reserves was maintained at the 35 U.S.$/oz., but in the open market, gold was allowed to float freely to seek its own price level. Finally, in 1971 President Nixon engineered the suspension of the Bretton Woods Agreement and pieces of paper, backed only by the full faith of the government became the new currency standard. Not much of a standard if one thinks about it. This is all good and well as long as the people maintain their faith in these small pieces of paper. As any tourist can attest, shirts, meals, just about anything can now be purchased in all corners of the world by paying in dollar currency. In Asia, South America, Russia, just about everywhere, prices to tourists are most often quoted in U.S. dollars, not rubles, rupies, and not barts, just plain dollars.

With the stroke of a pen, gold ceased its importance as an international payment instrument and became a commodity like wheat, pork bellies, and soy beans. Black gold, oil, has become a much more important commodity in the world markets than gold itself.

Newspapers were full of ads of gold coin investments such as a thousand dollars to control $10,000 worth of bullion, or a rare gold coin having acquired a tenfold increase in value over the past x-years. This declared gold to be a good and sound investment. Now, with gold prices having been stagnant for many years, investors have lost interest and many precious metal and coin shops have closed their doors. But they will resurrect again as soon gold prices tend to go through the roof again. Perhaps in two, ten, or twenty years, who knows? In normal times, gold coins are collector items like stamps, paintings, and art objects. There is nothing wrong with someone collecting coins as a hobby. Let's set the record straight. Gold, under the current low interest rate environment, is not an investment in our context and is highly speculative in nature. Our investment philosophy and strategy has no room for such risky ventures.

Much has been written about gold being the preferred hard asset during periods of a national or world crisis. This is well-documented and cannot be overlooked. During a severe political, monetary, or military crisis, investors have always fled to the safe haven provided by gold. Many nations still hold physical gold reserves in the treasury vaults as collateral for their paper currency. Today, many politicians in the U.S., Switzerland, and other nations would like their government to divest itself from this gold backing with the somehow short-sighted argument that gold doesn't generate a fixed return. It just sits there in the vaults of the national banks, so they say.

However, there have been, and likely will again be times in the future, when crisis conditions cause the value of gold to dramatically rise. It is in these abnormal times of crisis that we will consider gold funds for our portfolio. Because we are not capable of forecasting the future, we will not, repeat not, continuously maintain a certain percentage of gold funds in our portfolio. We don't structure our portfolio around things that might or might not happen in the future. We will add this type of investment when conclusive evidence is at hand that a crisis exists. Not before. We also immediately sell all our gold fund holdings as soon as evidence confirms that a crisis has passed.

11.2 Indicators for Gold Fund Investments

What then are these crisis conditions that will let us consider the possibility of adding gold funds to our portfolio? The indicators for gold investments to consider are

1. Inflation rates exceed 12%,
2. The domestic currency is in a free fall, i.e., loses more than 30% in a period of 6 months,
3. War breaks out among one or more of the industrialized nations,
4. A domestic banking crisis erupts with massive and panic-like withdrawals of deposits by the public,
5. Spot crude oil prices rising more than 50% in a 6 month period.

There is no need for a Fuzzy Logic path for gold fund investments. Everybody will know when such a crisis exists for it will be the major topic of discussion in the media and in the streets. We do not invest in gold funds in anticipation of a crisis; we invest in these instruments when the crisis is real and acute. Gold fund investments then, for us, mean purchases of funds that invest in gold mining companies. In such instances we will substantially reduce other equity fund holdings and certainly all bond funds by switching into gold funds. Depending on the severity of the crisis, we might even opt to stay away from gold mining stock funds and choose instead to convert our paper assets into physical gold bullion. Unfortunately, everybody else will do the same thing, compelling governments to impose restrictions on how much gold an individual investor will be allowed to hold. Let's hope we will never come to this crossroad. Worry about it? No way. We address these situation when they happen, but we surely don't want to lose sleep over it.

As soon as the end of a crisis has been confirmed, we will sell all our gold-related assets and again enter the world of mutual fund investments as we know it today.

12 The Fuzzy Logic Keys

To this stage we have discussed how an investor can use the Fuzzy Logic concept to determine the current basic condition for any type of investment. We have explained what indicators to use, but little mention has been made of how these indicators interact with our investment strategy. The following discussion will concentrate on this aspect.

At frequent intervals, we check the media to determine the current trend of our key indicators. The important word here is *current*. Do not be misled by forecasts of market experts and economists. The question of where the markets will be in six or eight months will always be dominant in the minds of investors.

For us, of prime importance will only be the question if current conditions represent an opportunity to buy, sell, or hold a given fund investment. As mutual fund investors we don't have to concern ourselves with such things as the balance sheet of a corporation, how well the company is managed, and how much profit the company expects to make in the next year. We let the fund manager do the raisin picking, hoping that he will load the fund portfolio with winners and with a minimum of poorly performing assets. For this expertise we pay him a management fee. A fund manager with an acceptable long term performance record will attract our intention and our money. We will not entrust our own money to a short term star performer or mediocre manager.

We now have finally arrived at the stage where we can discuss the Fuzzy Logic Keys, the solutions to given conditions that will be used to guide us in our fund investment decisions. The Fuzzy Logic diagrams each have a set of conditions numbered from 1 to 27 or, in the case of Bond or Foreign fund investments, from 1 to 9. Once this number is known, we check the Fuzzy Logic Keys to determine the possible actions we could take to adjust our portfolio holdings if needed. The key opportunities described herein are guidelines only and should not be used as the sole determining factor to change the portfolio structure. Before any action is taken, an investor must ascertain that the portfolio risk matches his own personal risk level.

Axiom IX

Portfolio adjustments are only made when the current portfolio risk deviates more than 3 points from an investor's own risk level or when the Fuzzy Logic Keys indicate a strong buy or sell signal.

One must guard one's self from switching in and out of funds on short notice without having done the required and proper analysis first. We don't want to become market timers; it doesn't fit our strategy. We don't want to be frequent fund switchers every time our Fuzzy Logic diagrams and action keys indicate a buy or sell opportunity. So-called market timers, or kangaroo investors, jumping from one fund to another at frequent intervals, very seldom show good consistent long term performance records. More often than not, their returns are worse than if they had followed a buy-and-hold strategy.

In our investment strategy, we are guided by well established indicators and fundamental principles. Although most of these principles have been discussed earlier, a short summary is given here again for a better overview of our concept.

12.1 Gross National Product

(Used for the assessment of the current economic condition.)

Fuzzy Logic Path:

HIGH = increase more than +0.4% in a month's time
OK = increase/decrease less than 0.4%
LOW = decrease more than –0.4% in a month's time

A recession is defined as a period in which the Gross National Product (GNP) declined two consecutive quarters. Likewise, a recovery of economic conditions is defined when the GNP increased for two consecutive quarters. When the GNP growth rate is larger than 3.8% and still increasing, it could be a signal of inflationary periods ahead. When the GNP has remained at levels below 1% and is declining, it could signal deflationary periods ahead.

12.2 Inflation Rate

(Used for the assessment of the current economic condition.)

Fuzzy Logic Path:
- HIGH = increase more than +0.4% in a month's time
- OK = increase/decrease less than 0.4%
- LOW = decrease more than –0.4% in a month's time

The best surrogate to measure inflation rates is the Consumer Price Index (CPI). Annual inflation rates higher than 12% could signal hyper-inflationary times ahead. Deflationary times are defined as periods when inflation rates declined for two consecutive quarters.

12.3 Interest Rates (Fed Fund and 10-Year Treasury Bond Yield)

(Used for assessment of current economic, equity and bond fund conditions.)

Fuzzy Logic Path:
- HIGH = increase more than +0.4% in a month's time
- OK = increase/decrease less than 0.4%
- LOW = decrease more than –0.4% in a month's time

Changes in the Central Bank's fund rate are signals of prevailing government monetary policies. For example, a U.S. investor noting an increase in Federal fund rate signals the Federal Reserve Bank is tightening, and a decrease in rates signals a loosening of the money supply. Tighter money policies reflect themselves in increasing interest rates. Conversely, more money supplied by the Central Bank results in lower interest rates. The equity and bond markets usually react favorably to a lowering of interest rates. The downside to lower interest rates is that money market yields and the value of the domestic currency will go down when compatible rates in other countries are much higher. Naturally, monthly distributions on the money market and bond funds held in the portfolio will also be lower. This is important for investors holding the major portion of their portfolio in these income-producing assets. In very low interest rate environments, the real possibility exists of the portfolio not delivering the minimum required return, as discussed in this book, to offset prevailing inflation and tax rates.

12.4 Price-to-Earnings (P/E) Ratio

(Used for the assessment of the current equity market conditions.)

Fuzzy Logic Path:
HIGH = when P/E higher than 18
OK = P/E between 12 and 18
LOW = when P/E lower than 12

When an equity market index's Price-to-Earnings Ratio is higher than 18, it is considered an indication of the stock market being overvalued by investors. P/E lower than 12 indicates an undervaluation. The above guidelines apply only to a market index as a whole, such as the Dow Jones Industrials, and not for individual securities. At a tail end of a recession, P/Es tend to be high and are thus to be considered in our Fuzzy Logic concept as OK.

12.5 The Gap from the Moving Average

(Used for the assessment of the current equity market conditions.)

Fuzzy Logic Path:
HIGH = when the gap is higher than +5.0
OK = gap between –3.0 and +5.0
LOW = when the gap is lower than –3.0

The gap is the percent difference between the current market index level and our weighted moving average (WMA). Since the weighted moving average is the trend line for the market index, the gap is thus an indicator of the market moving ahead or behind the trend.

High positive gaps indicate excessive bullishness, while large negative gaps signal excessive bearishness by investors.

12.6 Currency Exchange Rate

(Used for the assessment of the current foreign fund investment conditions.)

Fuzzy Logic Path:
HIGH = Investor's own currency trend is up
OK = Investor's own currency exchange rate is stable
LOW = Investor's own currency trend is lower

Attractive opportunities exist for foreign fund investments when our own currency exchange rate is in a declining trend against the foreign currencies of the countries where we are interested to invest. Conversely, an increasing trend in our own currency exchange rate signals less opportune times for foreign investments.

12.7 Foreign Equity or Bond Market Trend

(Used for the assessment of current foreign fund investment conditions.)

Fuzzy Logic Path:
HIGH = foreign market trend up
OK = foreign market trend stable
LOW = foreign market trend down

Market trends in foreign countries of interest to us are usually reported in the news media on a regular basis.

12.8 The Fuzzy Logic Keys

We have finally arrived at the point where the Fuzzy Logic principle can be implemented. Once we have, by our analysis, established the Fuzzy Logic

Condition Number, we look up the Fuzzy Logic Keys that apply to each individual condition. In short, once we have found the appropriate condition number for each type of fund investment, we can then use the Fuzzy Logic Keys as guidelines to structure our portfolio mix to our specific needs and risk levels. The keys indicate to us when to buy, sell, or hold specific securities in accordance to our investment concept. These keys are shown in the following Tables 12.1 to 12.3.

The herein proposed investment strategy allows for a systematic approach in matters of mutual fund investing. We believe the herein proposed risk, asset allocation, and investment strategy represents a new approach and a time-saving method for the typical mutual fund investor. This method might have broader applications for management of pension funds and investment advisors. Unquestionably, others will be able to improve on the selection of indicators to make the concept of investing by Fuzzy Logic even more powerful. This is quite acceptable because we do not set claim that this method is the ultimate answer to all the confusion that reigns in the investment community. There is nothing in the world that is not in need of improvements; nothing and nobody is perfect. We must accept the rapid changes in the world that have taken place with the advent of the computer age. If a car engine can be run smoother and more efficiently by the use of a tiny Fuzzy Logic chip, then it should also be possible to fine-tune investment decisions as well with the Fuzzy Logic concept.

Since 1971, when the author first developed the Fuzzy Logic Control concept, improvements have been made in the selection of the variables and in electronic control, but the basic three-dimensional TOO HIGH-OK-TOO LOW concept of Fuzzy Control has not changed. We expect the same thing will happen to the investment strategy discussed in this book.

12.9 Summary of the Investment by Fuzzy Logic Concept

Our investment concept has now been completely explained. For a quick overview, we summarize now the basic steps as discussed throughout the text of this book. The steps a mutual fund investor will take, applying our method, are also shown in Figure 12.1. This flowchart gives a quick overview of the

**Table 12.1 Fuzzy Logic Keys
Opportunities for Equity Fund Investments**

Key Number	*Economy*	*P/E*	*Gap*	*Guidelines*
1	LOWER	HIGH	HIGH	STRONG SELL
2	LOWER	HIGH	OK	SELL
3	LOWER	HIGH	LOW	HOLD
4	LOWER	OK	HIGH	SELL
5	LOWER	OK	OK	HOLD
6	LOWER	OK	LOW	HOLD
7	LOWER	LOW	HIGH	SELL
8	LOWER	LOW	OK	HOLD
9	LOWER	LOW	LOW	HOLD
10	OK	HIGH	HIGH	SELL
11	OK	HIGH	OK	SELL
12	OK	HIGH	LOW	HOLD
13	OK	OK	HIGH	HOLD
14	OK	OK	OK	HOLD
15	OK	OK	LOW	BUY
16	OK	LOW	HIGH	HOLD
17	OK	LOW	OK	BUY
18	OK	LOW	LOW	BUY
19	HIGHER	HIGH	HIGH	SELL
20	HIGHER	HIGH	OK	HOLD
21	HIGHER	HIGH	LOW	HOLD
22	HIGHER	OK	HIGH	SELL
23	HIGHER	OK	OK	HOLD
24	HIGHER	OK	LOW	BUY
25	HIGHER	LOW	HIGH	HOLD
26	HIGHER	LOW	OK	BUY
27	HIGHER	LOW	LOW	STRONG BUY

Guidelines should be used only in conjunction with the investor's own Risk Level and Portfolio Asset Allocation Risk.

thought and analytical process a mutual fund investor will follow when applying our investment concept.

Table 12.2 Fuzzy Logic Keys
Opportunities for Bond Fund Investments

Key Number	*Interest Rate Trend*	*Inflation Rate Trend*	*Guidelines*
1	LOWER	HIGHER	HOLD
2	LOWER	OK	BUY
3	LOWER	LOWER	STRONG BUY
4	OK	HIGHER	SELL
5	OK	OK	HOLD
6	OK	LOWER	BUY
7	HIGHER	HIGHER	STRONG SELL
8	HIGHER	OK	SELL
9	HIGHER	LOWER	SELL

Guidelines should be used only in conjunction with the investor's own Risk Level and Portfolio Asset Allocation Risk.

Table 12.3 Fuzzy Logic Keys
Opportunities for Foreign Fund Investment

Key Number	*Foreign Market Trend*	*Foreign Currency Trend*	*Guidelines*
1	LOWER	LOWER	SELL
2	LOWER	OK	SELL
3	LOWER	HIGHER	HOLD
4	OK	LOWER	SELL
5	OK	OK	HOLD
6	OK	HIGHER	BUY
7	HIGHER	LOWER	SELL
8	HIGHER	OK	BUY
9	HIGHER	HIGHER	BUY

Guidelines should be used only in conjunction with the investor's own Risk Level and Portfolio Asset Allocation Risk.

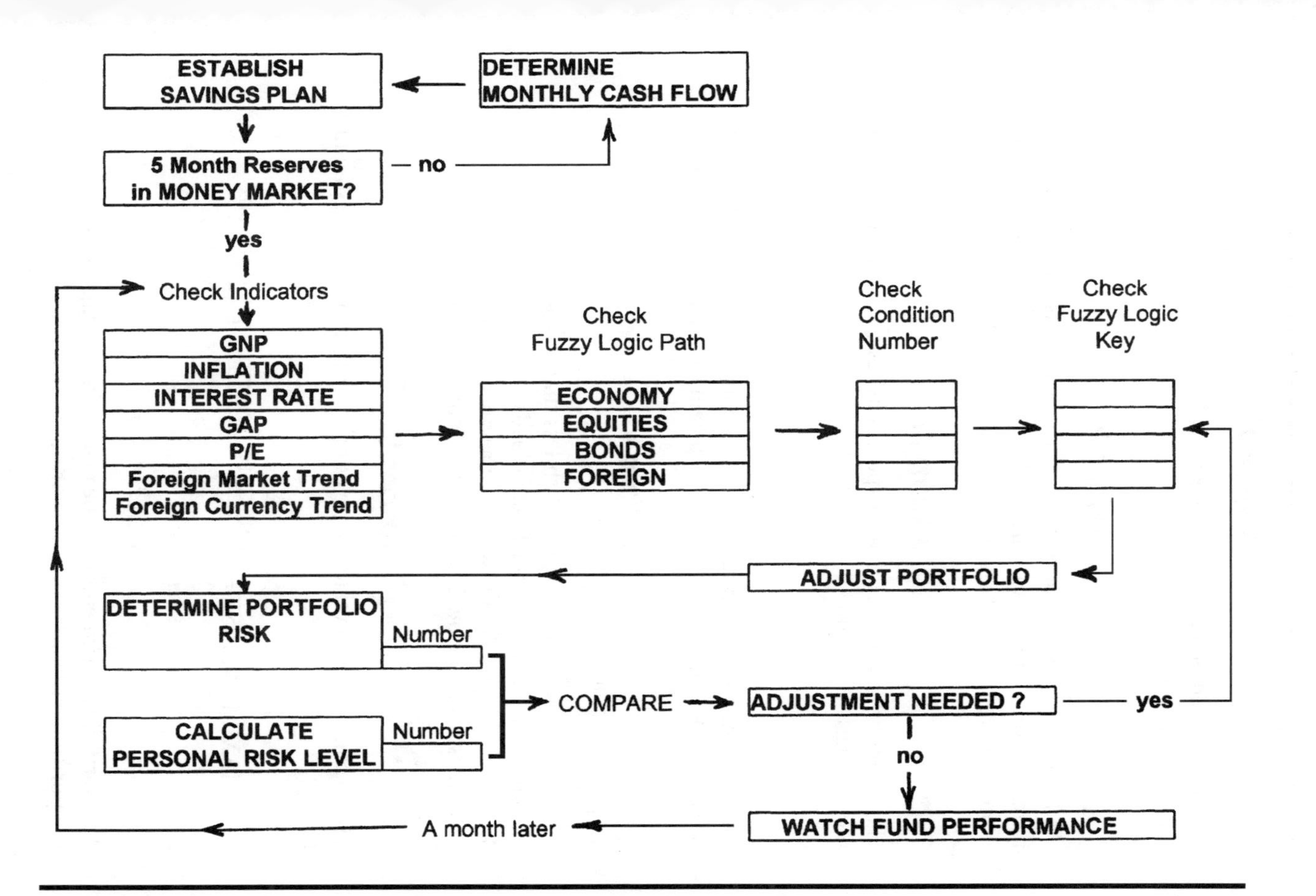

Figure 12.1 Flowchart for Mutual Fund Investments by Fuzzy Logic

Axiom I

Commit a fixed percentage of your monthly income for savings and make this a long term commitment.

Axiom II

With the exception of mortgage and car payments, pay cash for all purchases or, if credit cards are used, pay off the balance within one month to avoid interest charges.

Axiom III

Have 5 month's worth of living expenses as reserves at all times in a money market account.

Axiom IV

The minimum simple annual return of an investment portfolio must be at least 1% higher than the sum of current inflation rates plus taxes.

Axiom V

1. Invest only in open-end mutual funds
2. Buy no-load funds or funds with less than a 3% load
3. Buy funds with no 12b-1 hidden loads
4. Preferably buy funds with management fees below 1.2% for domestic and below 2% for foreign funds.

Axiom VI

The maximum risk level for an investor's portfolio as a whole should be

$$\text{MAXIMUM RISK} = \frac{\text{years to retirement}}{2} + \frac{\text{portfolio size}}{50{,}000}$$

AXIOM VII

Any portfolio adjustments must be made with the objectives to match the portfolio risks with the investor's personal risk level.

Axiom VIII

Do not attempt to time the markets by switching in and out of funds whenever a single indicator flashes a buy or sell signal.

Axiom IX

Portfolio adjustments are only made when the current portfolio risk deviates more than 3 points from an investor's own risk level or when the Fuzzy Logic Keys indicate a strong buy or strong sell signal.

---

13 Keeping Track of Investments

One of the unpleasant tasks every investor is confronted with is to keep order in his investment files and to frequently update his portfolio value. It doesn't make sense to have a drawer full of investment reports, fund brochures, securities, and other related matter without periodically cleaning it out. We also want to periodically check how our investments are doing to ascertain that our investment goals are still being met. This all comes with the turf.

Keeping all investments within one single institutional investment house, one that offers all the types of funds we require for our portfolio, will greatly reduce the amount of bookkeeping chores. Any portfolio changes made during a year or even a quarter will be neatly reported in a single statement. An investor who had to deal with a smorgasbord of investment reports from different fund companies in the past can attest to the fact how cumbersome this paperwork is when tax filing time comes around. We don't want to become bookkeeping worms and spend an inordinant amount of time as accountants. Since our investment style doesn't call for frequent changes in our portfolio structure, perhaps once or twice a year, and staying with one investment house does give us a certain relief from these chores.

For this purpose we make use of a spreadsheet as shown in Table 13.1 where all our investments are summarized on one single sheet of paper. This computer program not only calculates the current asset allocation mix but updates the portfolio anytime the Net Asset Value (NAV) or the number of shares held is changed. It also shows the portfolio's as well as the individual fund's year-to-date performance. Included in this form is also the calculation of the portfolio risk level and the asset allocation — two important variables

Table 13.1 Investment Portfolio Performance and Fund Watch

Date:								Maximum Risk Level:		27
Name:								Min. Return Needed:		5.8
Fund	*Account*	*Type*	*Current*			*Beginning of Year*				*% Chg*
Name	*Number*	*Fund*	*Shares*	*NAV*	*Value*	*Shares*	*NAV*	*Value*		*Ytd*
a		Money Market	22115	1.00	22115	21566	1.00	21566		2.5
b		Money Market	8755	1.00	8755	8551	1.00	8551		2.4
c		Aggressive Growth	1234.3	24.35	30055	1176.4	22.15	26057		15.3
d		Aggressive Growth	845.7	31.10	26301	845.7	26.35	22284		18.0
e		Aggressive Growth	0	56.22	0	0	51.66	0		
f		Equity-Income	125.6	22.10	2776	111.2	22.10	2458		12.9
g		Equity-Income	0	18.55	0	0	16.22	0		
h		Equity-Income	1225.3	29.65	36330	1201.3	27.15	32615		11.4
i		Intermediate Bond	1536.2	11.15	17129	1475.6	10.98	16202		5.7
k		Intermediate Bond	0	9.88	0	0	9.75	0		
l		High Income Bond	658.2	10.22	6727	622.35	10.18	6336		6.2
m		Foreign	196.2	35.16	6898	196.2	33.68	6608		4.4
n		Foreign	0	26.22	0	0	21.6	0		
				Total:	157086		Total:	142677		10.1

Asset Allocation	Total	Percent
Money Market	30870	19.7
Aggressive Growth	56356	36.0
Equity-Income	39106	24.9
Interm./Long Bonds	17129	10.9
High Income Bond	6727	4.3
Foreign	6563	4.2
Total	156751	100.0

Portfolio Risk Level: 25

that will automatically be updated whenever there is a change in any of the funds listed. Table 13.1 is only an example and can easily be modified to suit the reader's own individual needs and preferences. A once-a-month updating of this table seems to be a suitable interval for our purposes.

13.1 The Fund's Net Asset Value

The Net Asset Value (NAV) of a specific fund is updated every day and published daily in most newspapers. It can also be downloaded from the Internet by visiting the fund company's homepage. The calculations employed by the fund companies are relatively simple. They add up all the assets in the fund and subtract the liabilities to arrive at the Net Asset Value for the fund as a whole. They then divide this sum by the number of shares outstanding to obtain the Net Asset Value per share.

For example,

Total Assets:	
(value of investments and receivables)	= 98,000,000
Total Liabilities: (payables)	–1,200,000
Fund Net Assets:	= 96,800,000
Number of outstanding shares: **(3,200,000)**	
Net Asset Value Per Share	= 30.25

Again, we have left out the fund names in this spreadsheet to be impartial to the many existing fund families. Values for current NAVs for each fund can be found in the financial section of daily papers or can be downloaded directly from the Internet. Changes in the number of shares held, due to distributions, sales, or purchases are entered from statements received from the fund company.

13.2 The Base Costs

The major difficulty in keeping this form up to date is when a change is made to any of the fund holdings. At such times the NAV, the number of shares, and the beginning of the year figures have to be corrected to reflect the changes in base costs. This is important for year-end tax filing purposes and to show a true portfolio performance on a year-to-date basis.

Many fund investors fail to make these adjustments in the base costs and thus could possibly report gains to the tax authority that are higher than they really are. As good citizens we pay every cent of taxes that lawfully belongs to the government but — not one cent more. Once in a while tax authorities send us a friendly note with a check, reminding us that we have overpaid our taxes on our investment profits because of a miscalculation on our part. But this is more the exception than the norm. More often the note tells us to pay some more.

Let's illustrate this bookkeeping chore in detail by the following example:

Fund X		*Number of Shares*	*NAV*	*Total Value*	*Cost Basis*
January 1	Beginning of Year	566.45	21.76	12325.95	12325.95
July 8	Interest	18.55	19.88	368.77	368.77
Dec. 21	Capital Gains	46.38	23.15	1073.70	1073.70
Dec. 31	End of Year	631.38	23.56	14875.31	13768.42

For the preceding example, the wrong way to calculate gains (for tax purposes) would be to subtract the first of the year from the end of the year value: (14875.31 – 12325.95) = 2549.36 gain.

The correct way to calculate gains (for tax purposes) is to subtract the first of the year from the end of the year cost basis as shown on the following: (13768.42 – 12325.95) = 1106.89 gain.

Since both interest and capital gain distributions during the year are taxable, the total cost of the investment has to be raised by this declared amount. Doing it the wrong way would mean paying taxes twice. Over a period of say twenty years, such overpayment of taxes could amount to a sizable sum by itself. It therefore pays to devote the necessary time once a year to do these bookkeeping chores; it is an absolute requirement even if the portfolio size is still small.

Another method applies for tax deferred investments such as pension plans, IRAs, and KEOGH plans. Here, both the interest and capital gains are not taxable until the money is withdrawn from the account. Each withdrawal will then be taxed as income, hopefully at a lower tax rate until the entire portfolio's capital is used up. One might get the impression that calculating the cost basis once a year is not necessary with tax deferred accounts. Not so. Circumstances might arise where this information is still required by tax authorities in the future; therefore, it makes sense to still do these calculations on a timely basis.

13.3 Implications of Fund Distributions

Investors must watch out for the distribution dates because these too carry a tax liability. On this day, the NAV of the shares drops by the amount of the distributed interest and capital gains, and the number of shares increase an equivalent amount when these distributions are reinvested in form of new shares. This is illustrated in the following:

	Before Distribution	*After Distribution*
Net Asset Value Per Share	16.55	16.10
Distribution Per Share	0.45	0.0
Total Shares Held	476.00	489.30
Total Value	7877.80	7877.80

The total value of the investment has not changed but the number of shares and the NAV have. Since distributions are taxable, an investor is well-advised to wait until after the distribution date to invest in any fund that is currently taxable.

Proper timing for buying and selling funds in a taxable account becomes extremely important in years where markets show extreme volatility as was the case in 1998. Assume a given fund has gone up 24% in the first six months of the year and the fund manager sold most of his holdings near the market peak which at first sight looks like a stroke of a genius. Then, in the remaining month of the year, the fund lost 36% giving the fund a net loss of –12% for the entire year. Now comes the hammer blow. Since the fund took profits near the peak of the market, its capital gains and dividend distributions in December and the tax liability for its investors will be quite sizable. In short, the fund not only lost money but left its investors with a high tax bill to boot.

This emphasizes the need for investors to be aware of the distribution dates of the funds they hold. He should also keep a close eye on the turn-over rate of the funds. A fund, with a turn-over rate of more than 100% is likely to provide its investors with a huge tax liability in a volatile market year as described above. Here again the golden rule prevails: buy after the distribution date, not before.

It doesn't make any sense to pay taxes when this can be legally avoided or delayed. In the first example, an investor planning to invest 10,000 into Fund "X" would save himself 76.13 in taxes (at a tax rate of 28%) if he waits until after the distribution date to make the purchase. 76.13 could mean one more screwdriver for the government, whereas, in your pocket, it could mean a nice evening out. OK — it doesn't get you into the opera but a good movie, a Mexican plate at El Montezuma, and two drinks can do, too.

Investors with a computer of their own at home can simplify this task of record keeping by making use of some excellent software specifically designed to manage money and/or investments. The cost for such software is well spent, they are relatively cheap (considering the benefits), and get better all the time.

13.4 The Tax-Sheltered Portion of the Portfolio

The importance for an investor to have a clear cut investment strategy to acquire financial independence has now become apparent. There is a need to do something most people are reluctant to accept, namely, start saving for retirement at an early age. It doesn't do any good to wait until one is in their fifties to think and plan about this important aspect of life. By that time it might be too late to achieve the goals. Waiting too long to establish a savings discipline might force someone into risky and speculative investments to quote: "catch up with other investors" who have already acquired a sizeable nest egg. Participating in speculative ventures and the urge to get rich quick can easily lead to financial ruin.

The major portion of a person's net worth is often the equity in the home they own and the amount of savings in tax-sheltered retirement plans. This could be in form of a self-directed IRA (Individual Retirement Account), tax deferred annuity, or company pension plan. Most employees drawing a salary or income through self-employment can qualify for such tax deferred plans. These plans are one of the very few things lawmakers have designed to benefit the middle class, i.e., the wage earners. Contributions toward these plans are deductable from taxable income. Interest earned and capital gains are not taxable until the money is withdrawn at retirement age.

An investor can obtain extra benefits when a company he works for sponsors a pension/profit sharing plan that contains provisions for the employer to match part of the employee's own contribution. Whenever such a plan is offered, an employee should take full advantage and contribute the maximum amount permissible into such a plan. Unfortunately, a vast number of employees don't. Where else can an investor get a guaranteed return of 25 to 100% in the first year? The only caveat of these plans is that an employee has to stay on his job for a given length of time, typically 5–10 years, before he is fully vested in the employer's portion in the plan. (The employees own contributions are, by law, always fully vested from the beginning.) In some countries, laws have been enacted to allow an employee to transfer the total pension amount over to his new employer when he changes jobs. This prevents "shady" employers from rotating employees for the purpose of manipulating pension benefit liabilities. These laws are also designed to stop corporations from terminating employees just prior to the time the employee becomes fully vested in the plan. Good loyal employees are the best asset a corporation can have. This loyalty does not come automatically; it has to be earned. Employers who have recognized these facts are usually also those that have the least amount of worker unhappiness on the job.

13.5 Tax Liabilities

In some countries, laws have been changed to significantly limit the tax deductibility of pension fund contributions. This is a clear mistake and another reflection of prevailing shortsighteness at government levels. Taking the incentive away for savings by employees and at the same time allowing corporations scores of tax deductions and loop holes just doesn't make any sense.

When retirement age is near, any investor with a sizable portfolio should get tax advice from a reputable professional. By this we don't mean advice from an investment house that sells investment products but advice from a bonafide tax expert. This is especially true for investors who contemplate early retirement before age 59 1/2 or investors who approach age 70. With few exeptions the law imposes a 10% penalty on any pension fund asset withdrawn before age 59 1/2. This penalty is in addition to the normal tax that has to be paid on the amounts withdrawn. An even higher penalty is imposed when an investor fails to draw the specified minimum amount from the pension plan after he has reached age 70 1/2. The first withdrawal must be made before April 1 after you have reached age 70 1/2. Missing the first withdrawal by just one month might get you and your portfolio into deep, deep trouble. Tax authorities could then wipe out 50% of the entire account amount. Not a very pleasant thought. So be prepared. Get in touch with a reputable tax advisor who knows all these pitfalls and do this well before you reach this critical age.

Suppose an investor, age 70, has an account worth 500,000. Actuarian tables might indicate that the man has a normal life expectancy to age 82.

Thus the tax authority dictates that these 500,000 have to be withdrawn at an annual rate of

$$\frac{\text{Principal}}{\text{life expectancy} - \text{age}} = \frac{500,000}{82-70} = 43,478 \text{ minimum per year}$$

When the investor withdraws only a portion of the calculated amount, the difference between the two will incur a penalty charge of 50% in the U.S.! Be on guard against these tax traps. Another example: for every liter or gallon of gasoline we buy, governments tax us from 20 to 35%, money to be used for road construction (or to drill a tunnel through the Alps). And what do they do when they have completed the work? Yes, charge us tolls to use these roads or tunnels.

13.6 Estate Planning

There are many tax angles related to investments and personal assets, most of them too complicated for the average investor to understand. As soon as you have accummulated a sizable estate, irrespective of age, think about the possibility of what could happen to these assets once you are not around anymore. This is not a morbid thought. None of us has a guarantee that we will still be alive tomorrow. We save our heirs a lot of headaches and taxes if we take the proper steps to prepare for such an eventuality. Keep in mind that most governments have a cap on how much of an estate can be tax-free passed on to the heirs. In the U.S. this currently amounts to $600,000 although Congress is contemplating raising this to one million. In other words, everything over $600,000 becomes taxable. So, after you leave an estate of say 1.6 million (that includes your pension, IRA, home, and proceeds from life insurances) your survivors will be clobbered by the tax authorities for the one million that will be taxable.

First and foremost, get legal and expert advice on this matter. Also get in touch with the mutual fund company for guidance; they usually have in-house experts that can steer you in the right direction. These experts might be able to set up a trust that will spare your surviving spouse hefty estate taxes. But, don't forget, when your spouse leaves this earth, too, the estate will then become a tax liability for the children. Estate taxes are thus only deferred; sooner or later the tax man will get what he considers his due. Isn't it ironic? First you get taxed on your income while you still work, then after you have saved some of your money, your savings become taxable, and last, when you die, you are again taxed for what is left over.

The costs for getting legal advice on estate tax matters are minimal compared to the vast sums of money your heirs might have to pay to the government. You cannot leave this to your heirs and do nothing about it. You must act now. Get legal advice if your current net worth is more than 600,000, irrespective of your current age.

There are legal ways in form of trusts that can ease the tax burden. Certain trusts might allow you to transfer part of your assets to your children, even grandchildren, to save them the headaches of estate taxes when your spouse also passes away later. In the U.S., you can also give away, tax-free, $10,000 every year to your children, relatives, or friends to reduce your taxable estate. You can finance the college education of your grandchildren. There are many legal ways in which you can gradually distribute your "excess" wealth to others so the government will have less to take after you have taken the final voyage.

If at all possible, don't leave your assets to the estate. Be on guard against ill designed wills. It can be full of "potholes" for the heirs if not properly prepared.

If you decide to set up a trust in your lifetime, make sure its an irrevocable trust. Talk this out with your tax advisor (he must however know this business; there are a lot of them around who don't). Don't expect the nearest Internal Revenue Service to provide you with the best alternative answers to your questions. These people are trained *to collect* not *to save* you and your heirs money. There is nothing wrong in also discussing this with your spouse and children — it can save them a lot of headaches later. Tax laws applying to inheritances tend to be most lenient when the assets are bequeathed by a written and legal will to the surviving spouse. However, when the assets involve a large total sum, it pays to consult an attorney to inquire if setting up a trust might save considerable taxes for the spouse and the surviving children. Leaving all the assets to the spouse outright might incure estate taxes of 18% whereas, with a trust, the tax liability might be only in the neighborhood of 5%. For a 700,000 estate, the tax savings could thus amount to as much as 90,000.

When one leaves his IRA or tax deferred account to his spouse, it might even be possible for the surviving spouse to roll over the entire account into her or his own tax deferred account. But, again, check this out. There are so many questions and uncertainties — get these answered now, as soon as possible. And — have a lawyer set up a legal will; better too early than never.

Probating an estate is a cumbersome task especially when the executor of your estate doesn't have all the necessary information he needs. We don't want to leave a mess behind. Proper estate planning will ease the burden on our beneficiaries. We can make it easier on them and save them the nightmare of spending weeks or months to obtain information when we can do the same thing for them in a few hours.

It doesn't take much time to prepare a list of items that are of prime importance to the attorney, tax authority, and your beneficiaries. To help the reader set up such an information sheet, we have developed a table for this purpose. We are not far off the track when we state that anybody with more than $20,000 in assets, regardless of age, should have such a list safely tucked away some place in his home (not the safe deposit box whose access could be blocked for month). It is however just as important to let your primary beneficiary know where this information is stashed away.

The list, as shown in Table 13.2, is by no means complete. It must be viewed as a starting incentive and can be enlarged or altered in any way by the reader's preference.

13.7 Taking Early Retirement

Retirement — how delightful when one has the financial resources to enjoy it. Most people like to take early retirement to do things they never had time for during their working years like going on cruises, playing golf three times a week — doing work and chores when one feels like it and not when the employer mandates it. Taking early retirement requires two considerations: first there must be enough financial resources available to do so and, second, the retiree must be sure he has things to do to prevent him from becoming bored and lonesome. Many companies, forced for one reason or another to downsize the workforce, offer their older employees a bonus to entice them to take early retirement. Everyone confronted with such an offer must take a very hard look at his financial situation, must try to negotiate for a better settlement, and not take the employer's word at face value when told how great a deal his offer is. Let's face it; they want to get rid of older employees and one might just as well take a strong and tough stance in such situations.

There is a danger of underestimating the monthly income needs because of the widely held beliefs that one needs less money in retirement. Even magazines and investment experts advise that only 80% of the pre-retirement income will be needed because of the assumption that taxes, transportation,

Table 13.2 Estate Planning Worksheet

Date:

1. Name, address of estate owner
2. Location where will is stored
3. Names and addresses of dependents (spouse and children)
4. Names and addresses of beneficiaries
5. Name and address of attorney to contact
6. Type of will or trust set up
7. Name, address of independent executor of estate
8. Social Security number, taxpayer I.D.
9. Location where past tax return copies are stored
10. Municipality, city, and state where taxes have last been paid
11. Name and address of employer (from which wages, benefits are due)
12. List of debitors (those that owe you royalties, payments, money, etc.)
13. Names and addresses of renters living in your rental property
    Include in all: account number, name, address, & phone number
14. Bank accounts
15. Location of safe deposit box
16. Broker accounts
17. Mutual fund accounts
18. Life insurance
19. Car insurance
20. Home insurance
21. Liability insurance
22. Health insurance
23. List of items you still owe money on (installment plans, contracts)
24. Location of real estate owned
25. Location of deed and trust
26. Mortgage company name, address
27. If Renting: name & address of landlord and leasing agent
28. Appraised property value
29. List of all personal property incl. estimated resale value
30. List of all investments and current value
31. Name, address, phone number of all persons to contact
    Where appropriate: enclose copies of important documents

food and clothing costs will be less. We completely disagree with this theory. If an investor truly wants to enjoy retirement, he should have at least the same if not higher living costs than before retirement. We also don't agree with the standard procedures employed by investment advisors to project a life expectancy of 84 for women, 82 years for men. One better plan his retirement around the assumption that one lives at least to age 90.

Another consideration must be future inflation rates since these alone can severely reduce the purchasing power of the savings accumulated for retirement. Obviously, one should also not retire unless one is covered by a good medical insurance plan. Medicare by itself is not good enough and, more important for the investor taking early retirement, Medicare doesn't become effective until the normal retirement age is reached. Let's face it, what good does a half million portfolio do when it can be wiped out in large part by a major medical expense? We must be prepared for such eventualities in the twilight of our lives and the only protection provided for these emergencies is a rock-solid good medical insurance.

We have developed a spreadsheet that will enable an investor to determine if early retirement makes sense or if the investor would be better off to keep on working. The advantage of this spreadsheet is that it allows an investor to study variations in outcomes by selecting his own "what-if" situations. What happens when the inflation rate is 4% instead of 6%? What happens to our portfolio when we withdraw 4000 per month instead of 3500? The reader can select any kind of combinations of possible future events by changes in input variables. An example of this spreadsheet is given in Table 13.3 along with the formulas used.

The formulas used in this spreadsheet take consideration of the inflation rates, an important aspect that is often overlooked. When an investor can live today on 4000 a month, it certainly doesn't mean that he can live on the same monthly income 25 years in the future. Look back at what groceries, housing, or cars cost 30 years ago. At that time a family might have been able to get by with an income of 2000 a month; today the same amount of income might just hold this family above the poverty level. We thus have increased the living costs and portfolio withdrawal amounts every year by the inflation rate in these calculations, partly compensated by the cost of living increases in Social Security receipts.

The reader would have to decide on his own if he wants to draw down his nest egg completely by age 90 or if it would suit his objectives better to leave part of the principal intact. Each investor has his own reasons for this issue, but it would be more prudent to have a major portion still available at age 90. Who knows? We might be able to become centenarians. Don't forget: life begins at eighty. Once over eighty, one can be doing just about everything without offending people. Do something stupid at this age and

Table 13.3 Monthly Income Declining Portfolio

Input	
P = Principal Amount at Retirement	475000
i = assumed Annual Portfolio Return (%)	0.065
f = assumed Annual Inflation Rate (%)	0.03
t = assumed Annual Tax Rate (%)	0.11
x = Years Remaining until Age 59 1/2	2.5
y = Years Remaining until age 65	5
s = Soc. Sec. and Pension Benefits (Monthly)	2650
e = projected Monthly Living Costs	3700

	Annualized					
Year	*Start of Year Principal*	*Portfolio Return after Tax*	*Income Needed*	*Withdraw Penalty 10%*	*Soc.Sec Pension*	*Ending Principal*
1	475000	27478.75	44400	4440	0	453639
2	453639	26243	45732	4573	0	429577
3	429577	24851	47104	2355	0	404968
4	404968	23427	48517	0	0	379879
5	379879	21976	49973	0	0	351882
6	351882	20356	51472	0	31800	352567
7	352567	20396	53016	0	32754	352701
8	352701	20404	54606	0	33737	352235
9	352235	20377	56245	0	34749	351116
10	351116	20312	57932	0	35791	349287
11	349287	20206	59670	0	36865	346688
12	346688	20056	61460	0	37971	343255
13	343255	19857	63304	0	39110	338919
14	338919	19606	65203	0	40283	333605
15	333605	19299	67159	0	41492	327237
16	327237	18931	69174	0	42737	319731
17	319731	18496	71249	0	44019	310997
18	310997	17991	73386	0	45339	300941
19	300941	17409	75588	0	46699	289462
20	289462	16745	77856	0	48100	276452
21	276452	15993	80191	0	49543	261796
22	261796	15145	82597	0	51030	245374
23	245374	14195	85075	0	52561	227054
24	227054	13135	87627	0	54137	206699
25	206699	11958	90256	0	55761	184162
26	184162	10654	92964	0	57434	159287

Table 13.3 (continued) Monthly Income Declining Portfolio

27	159287	9215	95753	0	59157	131906
28	131906	7631	98625	0	60932	101844
29	101844	5892	101584	0	62760	68912
30	68912	3987	104632	0	64643	32910

people feel sorry for us and say we have gone senile. Do something ordinary and people will call us wise old men or women.

The first requirement for use of this spreadsheet is to establish our monthly income needs in retirement and, as mentioned, we must be generous in our expectations. A detailed study of these requirements can prevent heartaches and shortfalls in the future. Costs for travel, health insurance, and medication will have to be projected higher than what they are now. We also have to know the sum of liquid money (principal) that is available in our portfolio from which we can make the withdrawals. We do not include the equity in our house if we plan to live in it during our retirement years. If we plan to sell the home, we can add the net sale proceeds, but must make provisions in our monthly costs for the rent we will have to pay thereafter in our new location.

We also have to enter the number of years remaining until age 59 1/2 (when IRA withdrawal penalties stop) and age 62 or 65 when we will be eligible to collect Social Security payments. The final input needed is the amount of Social Security and pension plan payments we will receive.

Again we repeat, we absolutely must be convinced without any financial doubts that we want to retire early. Our life style will change then and we must expect that such an early retirement might not suit us for long. We might regret such a move after one or two years. If there is any doubt, we might be better off to keep on working and delay retirement. Unless we feel adventureous to start our own business or a brand new career. Like the welder who retired at age sixty and two years later had 400 people under him. His new career — cutting the lawns in a cemetery.

How many years will our portfolio last if we withdraw a given amount every year, inflation adjusted? For the reader who doesn't want to go through the elaborate work of setting up his own spreadsheet to find the answer to this question, we have developed Table 13.4. A quick glance at this table will enable him to roughly get an idea of how well (or how bad-off) he will be financially in retirement.

Table 13.4 Number of Years Until Portfolio is Depleted

Initial Monthly	*Portfolio Size*										
Withdrawal	*20000 0*	*25000 0*	*30000 0*	*35000 0*	*40000 0*	*45000 0*	*50000 0*	*55000 0*	*60000 0*	*65000 0*	*70000 0*
1000	24	>30	infin.	infin.	infin.	infin.	infin.	infin.	infin.	infin.	infin.
1500	14	18	24	>30	>30	>30	infin.	infin.	infin.	infin.	infin.
2000	10	13	16	20	24	28	>30	>30	infin.	infin.	infin.
2500	8	10	12	15	17	21	24	27	>30	>30	infin.
3000	6	8	9	11	13	16	18	21	24	27	>30
3500	5	6	8	10	11	13	15	17	19	22	24
4000	4	6	7	9	9	11	12	14	16	18	20
4500	4	5	6	8	8	10	11	13	13	16	17
5000	3	4	5	7	7	9	9	11	12	14	14
5500	3	4	5	6	6	8	8	10	10	12	13
6000	3	3	4	6	6	7	8	9	10	11	12
6500	2	3	4	5	5	7	7	9	9	10	11
7000	2	3	4	4	5	6	7	8	8	9	10

Assumptions: Annual income from portfolio investments = 5%
Monthly withdrawals increase every year by 2.5% due to inflation

14 Final Words

Remarkably, the savings rate in the U.S. is less than 6% of the average household income whereas it is closer to 20% in Japan and 25% in Switzerland. We must wean ourselves away from credit cards and installment loans and become more concerned about the future instead of living from payday to payday. Unfortunately, we cannot look at governments as role models, neither can we ask the financial community for sensible guidance in this respect.

14.1 Government Fiscal Policies

Governments, worldwide, are the worst offenders when it comes to financially irresponsible behavior. Periodically lifting the debt ceiling seems to be their only answer; they want us to believe that more debt is the only solution to get out of financial difficulties. They try to convince us that more debt will stimulate the economy and reduce unemployment. It's their easy way out of a situation that normally demands drastic measures. Any new government debt exceeding the rate of economic growth means loading the new debt on the shoulders of future generations.

Government fiscal policies might be, at best, instant and temporary problem solvers that however could lead to much larger problems in the future. The previous budget surplus in the U.S. was in the middle 1960s; it took over 30 years of red ink before the fiscal house was brought in order again. What sticks to our mind was the back slapping and glorification the President and most of the Senators and Congressmen heaped upon themselves when they were able to announce the budget deficit had been eliminated. Let the good times roll again. But, there is no reason to bask in glory. Having achieved a balanced budget means only that tax revenues matched government expenditures. The time to dance in the streets comes only after they have cut government waste and lowered the tax burden of the average citizen. That's what a true and meaningful balanced budget means.

Europeans are walking the same tightrope. Even the "strong" currency nations like Germany and Switzerland are running horrendous budget deficits, and like everywhere else, don't want to cut costs or raise taxes. Why? Because doing so could mean political ruin for the party in power. No politician wants to admit or acknowledge that ill-advised governmental financial programs can lead a country into ruin.

Going into debt in difficult times is only acceptable, according to sound financial principles, when the debt is paid back once conditions have improved. Few governments, if any, will do it. Instead they prefer to use the gadget ordinary citizens don't have: they let the Central Bank print fresh money even in relatively good and stable economic times.

Proclaiming the stock market as sound after it dropped 500 points in one day or saying the banking system is in good state, when everybody knows it is on the brink of collapse, doesn't help bolster the confidence of the average citizen.

Devastating as it is, governments often deflate the value of their own currency whenever they choose to ignore the truth that the piper sooner or later has to be paid. A devaluation of a currency is nothing else than a cheap ploy to rob the average saver of his hard earned money. Some people adhere to the myth that a little inflation is not bad because it makes house equities rise and allows people to pay off debt with cheaper money in the future. In other words, they use the same befuddled financial mentality that prevails among many government officials, CEOs, and bankers.

Do we want to reach a state wherein we would have to be paid twice a week to allow us to convert worthless paper money into hard assets before the printed money loses ten or twenty percent of its value within a few days? Impossible? Perhaps, but let us not forget that people in Argentina and Brazil thought the same thing twenty years ago. Likewise in Germany in the 1920s. But, it did happen and it could happen again, even in our own country of residence. Could the steady decline of some currencies in the past twenty years be a warning sign of worse things to come? We don't know; we don't dabble in predictions. It only makes us alert.

We don't agree with government officials proclaiming a lower value of the domestic currency is good because it will promote exports. Yes, domestic products will then be cheaper for overseas customers, but what is good for industry might not be good for the consumer. Imported goods will become more expensive on the domestic market. Any time politicians try to water down a currency's fall, it should be a warning sign for us of internal financial problems.

In a worst case scenario, someone could possibly retire with an apparent nice income from his investments and company pension plan and find out, three years later, monthly retirement benefit would be just enough to buy the necessary food and not much else. A frightful thought. Many would say this could never happen because industrialized nations are now coordinating interest and currency exchange rates in an orderly and controlled fashion. You bet. Germany, Japan, China, and the U.S. cooperating with each other? We must stop believing in fairy tales. Each will cooperate as long it is to their advantage.

In the past, many believed a stock market crash and deflation such as the one of 1929 could never happen again, believing such an event would be impossible today because trading rules are now in place to prevent such a calamity. Despite of all these assurances, the market experienced the October 1987 meltdown. Two to three years later, the New York and Chicago exchanges placed circuit breakers in effect that would be activated when the market changes a given number of points. Think about it — does this solve the volatility problems caused by speculative behavior of large financial institutions? Does it make a difference if the stock market tumbles 1000 points in one day or by the same amount in twenty days because of these so-called circuit breakers? Does anyone believe these Band-Aid solutions for index arbitrage and futures trading will be able to stop a catastrophic drop in the equity market when investors, all at once, within a month's time, decide to exit the market? It would make more sense to raise the margin requirements for options and futures trading to the same level as is required for equity trading to reduce excessive market volatilities.

The author has in front of him two items that support the many critiques we have employed in this book. One is a newspaper clipping from a market analyst who based his prediction upon a new cycle theory he pioneered. His forecast: by 1992 the Dow Jones Industrials Index will stand at a level of 1000 or lower. Looking at the Dow charts in this book we know what happened and how well this new theory worked. Unquestionably, there must have been quite a few who acted upon this extreme forecast and, in turn, might have missed one of the strongest bull market moves in history. We know better than to follow such advice. To this day, nobody has been able to make consistent correct forecasts on future market directions. Predictions for us are taboo; we only look at the present conditions and decide from these if a change is warranted for our investment structure.

The second item is a recent Swiss newspaper article discussing why the "Big Three" banks in Switzerland were able to report 20 to 34% higher profits for the year 1995 despite a significant downturn of the Swiss economy. Some of the reasons given were not surprising and are also not in question since they related to results of restructuring and attempts to increase reserves. However some of the other reasons given awakened our curiosity. They were

- increased income due to booming equity and trade markets,
- capital gains on securities held by the banks,
- increased deposits by small investors into savings accounts — with new fee structures for small investors (higher costs).

In plain language, by being paid only 3% interest on their accounts, savers have allowed the banks to reinvest this "cheap savings capital" for their own account into higher paying securities and — pocketing the interest difference for themselves without much risk exposure. But, there are also banks that invest this cheap savings capital in very risky ventures such as hedge funds or other derivative instruments to enhance their profits, with disastrous results when they bet on the wrong side of the markets. This immediately raises questions such as

- Why were these banks able to make such good profits from the booming securities markets for their own account and, in the same year reported mediocre to poor returns for most of their own mutual funds that in most cases didn't even match prevailing inflation rates?
- Why does the small investor have to pay higher management fees irrespective if the fund had a good or a bad performance?
- Why does a fund manager still collect an inordinately high salary even when his fund performed much worse than others in the same group?
- Would it not make more sense to set a fund manager's salary and the fund fees in accordance to performance as is done by many mutual funds sold in the U.S.?

One cannot fail to get the impression that banks are primarily interested in producing good performance for their own accounts. Performance for their small customer accounts appear not to be a primary objective. It is not our intention to single out the Swiss banks; we have only used this example to bring to the forefront a problem that we think exists worldwide in the investment community. Small investors who entrust their hard earned money to the banks for investment have a right to demand good performance. They also have a right for transparency and full disclosure of what happens to their money invested in mutual funds, the least of which would be a semi-annual report of the types of securities the fund is invested in and a full disclosure of all the deposit fees and fund management costs. The mutual fund industry in the United States has perhaps the best transparency in this respect. It is perhaps for this reason that in no other country in the world are there so many mutual fund investors, on a per capita basis, as in the U.S. Many foreign banks seem to have forgotten that customer satisfaction at all levels is the driving force for repeat business and loyality.

14.2 The Author's Internet Homepage

By now the reader might have decided to implement the herein described investment technique for the management of his own investment portfolio. Others might want to give this technique a "dry run" on their computer first to test the concept before committing real money. Then there are those apprehensive of the task and time involved to keep track of the various indicators and conditions upon which our Fuzzy Logic concept is built. To stay on top of current developments, as I have learned, requires updating the data at least once a month. Some view this as an interesting activity and are quite willing to devote the necessary time to it. But there will be others who for many legitimate reasons do not have the time available to do this research. Finally, there undoubtly will be some just interested in the kind of signals our Fuzzy Logic Keys are giving under present economic and market conditions. To all those, we offer a solution in form of our own Internet Homepage: http://ourworld.compuserve.com/homepages/peray

This homepage will be updated once a month to keep the reader informed of any changes that might have occurred from the previous month's Fuzzy Logic conditions (see Figures 14.1 and 14.2). The structure of this homepage is built on the same principles as this book, leading the reader step by step through the entire analytical process. Starting from the homepage, click on any of the links to get to the pages of interest. Your investment portfolio risk might not correspond to your own personal risk level and you thus contemplate making changes in the portfolio's asset allocation. Using the link to the Asset Allocation tables, select the portfolio mix that, under present

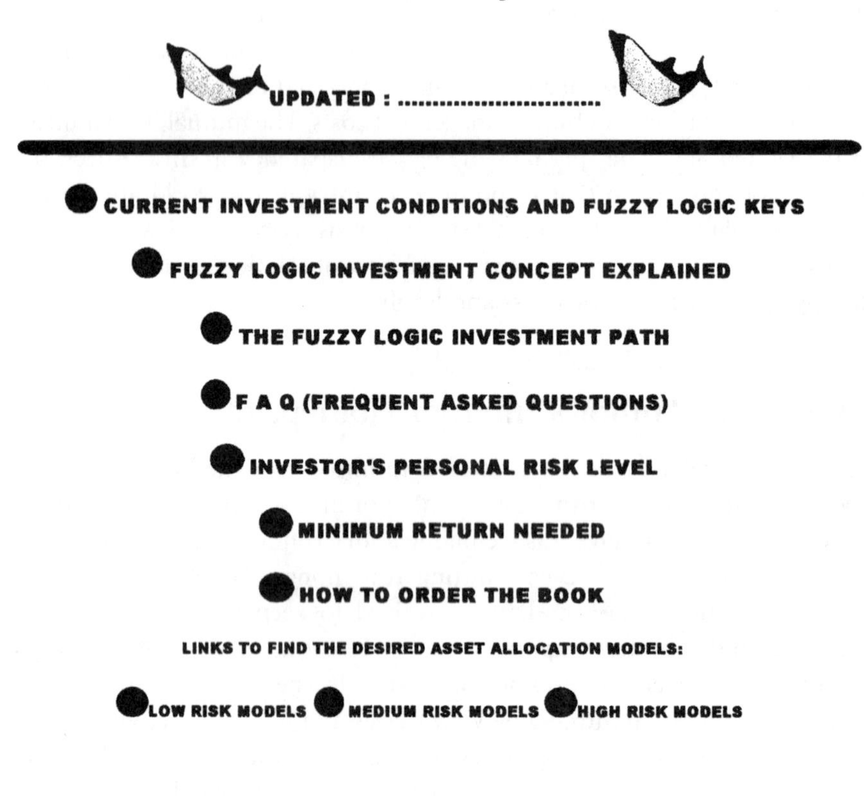

MUTUAL FUND INVESTMENTS BY FUZZY LOGIC©
Kurt E. Peray
UPDATED : ....
CURRENT INVESTMENT CONDITIONS AND FUZZY LOGIC KEYS
FUZZY LOGIC INVESTMENT CONCEPT EXPLAINED
THE FUZZY LOGIC INVESTMENT PATH
F A Q (FREQUENT ASKED QUESTIONS)
INVESTOR'S PERSONAL RISK LEVEL
MINIMUM RETURN NEEDED
HOW TO ORDER THE BOOK
LINKS TO FIND THE DESIRED ASSET ALLOCATION MODELS:
LOW RISK MODELS
MEDIUM RISK MODELS
HIGH RISK MODELS
All rights reserved
Our e-mail Address : peray@compuserve.com

Figure 14.1 Author's Internet Homepage

conditions, would best suit your own risk preference. Or, you might just want to check on the current status of the Fuzzy Logic Conditions be that for equity, bond, and/or foreign fund investments. In short, the author's homepage will give the reader the capability to stay informed of current developments that might have an impact on his investment portfolio.

Included in the author's homepage is also a link to order this book direct from the publisher. The reader can also use the author's e-mail link (peray@compuserve.com) to forward questions and comments to me related

UPDATED :

DOW JONES-IND		intrinsic value	

" Mutual Fund Investments by Fuzzy Logic "

	previous	current	Condition	Fuzzy Nr.	FUZZY KEY
ECONOMY					
GNP					
INFLATION RATE					
FED. FUNDS					
EQUITIES					
ECONOMIC CONDITION					
MARKET P/E RATIO					
GAP					
BONDS					
INFLATION RATE					
10yr TREASURY BOND YIELD					
FOREIGN INVESTMENTS	currency	currency trend	market trend		
Euro					
Germany					
Britain					
Switzerland					
Hong Kong					
Singapore					
China					

Note : Portfolio adjustments , using the Fuzzy Logic Keys, should only be made when the current Portfolio Risk doesn't match the Investor's own Risk Level.

GO BACK TO HOMEPAGE

Figure 14.2 Author's Internet Homepage, Monthly Conditions

to the topics discussed in this book. Although I cannot promise to answer all e-mail messages, I will make an effort to respond to as many as time permits.

15 Appendix

15.1 Formulas Used in Tables and Figures

The most important formulas used in this book are described here to allow the reader to construct his own computer programs using conventional spreadsheet software such as Excel and Works. The formulas are all stated in the style needed to enter into spreadsheet tables and graphs.

Table 1.1 Net Worth

Net Worth = Total Assets – Total Liabilities

Table 1.2 Monthly Cash Flow

Monthly Cash Flow = Total Income – Total Expenses

Table 1.3 Amount Needed for Retirement

P = 1/((0.05 * ((1.05) ^ y))/(x * (((1.05)^ y) – 1)))

let P = principal amount needed
x = annual withdrawals in retirement
y = years of expected retirement

Table 2.1 Compound Interest Factors

F = (1 + 0.01 * i) ^ n

let n = number of years
i = percent annual simple interest
F = compound interest factor

Tables 2.2 and 2.3 Portfolio Growth with Monthly Deposits

$$P_f = (d/i) * ((1 + i) \wedge (n + 1) - (1 + i))$$

let n = number of deposits
i = percent monthly simple interest
d = amount of monthly deposits
P_f = final value after *n* deposits

Table 2.4 Minimum Return Needed

$$M = (inf + 1)/(1 - 0.01 * t)$$

let t = tax rate (%)
inf = inflation (%)
M = minimum return needed

Table 3.4 Gains Needed to Break Even after a Loss

$$G = x/(100 - x) * 100$$

let x = initial loss
G = gains needed

Table 4.3 Mutual Funds Performance Index

for Equity Funds: $PI = 0.15 * a + 0.35 * b + 0.4 * c + 0.1 * d$

for Bond Funds: $PI = 0.30 * a + 0.20 * b + 0.25 * c + 0.25 * d$

let a = average annual total return in last 3 years
b = average annual total return in last 5 years
c = total return in *worst* year
d = total return in *best* year
PI = performance index

Table 5.1 Investor's Personal Risk Level

$$R = (n/2) + (a/50{,}000)$$

let n = number of years until retirement
a = portfolio size
R = investor's risk level

Tables 6.2, 6.3, and 6.4 Asset Allocation Models

$$x = a * (m - 1) + b * (-10) + c * (-6) + d * (-4) + e * (-10) + f * (-35)$$

$$y = a * (m + 1) + b * 24 + c * 20 + d * 15 + e * 19 + f * 48$$

$$PR = y - x$$

let a = % of portfolio in money market funds (expressed as decimal)
b = % of portfolio in aggressive equity funds (expressed as decimal)
c = % of portfolio in equity-income funds (expressed as decimal)
d = % of portfolio in interm. bond funds (expressed as decimal)
e = % of portfolio in high income funds (expressed as decimal)
f = % of portfolio in foreign funds (expressed as decimal)
m = current money market yield (percent)
x = portfolio *worst* index
y = portfolio *best* index
PR = total portfolio risk

Figures 8.2 to 8.12, Tables 8.1 to 8.11 Plotting Moving Averages and the Gaps of a Market Index

$$WMA = 0.128 * C + 0.872 * p$$

$$GAP = ((C/WMA) - 1) * 100$$

let WMA = current weighted moving average
C = current market index
p = previous week WMA
GAP = difference between the current market index and the WMA

Table 13.3 Monthly Income — Declining Portfolio

A = P(prev) (Note: in Year 1 = Principal at retirement)

$$B = (A * i) - (A * i * t)$$

$$C = C(prev) * (1 + f)$$

D = C * constant (0.1 if funds withdrawn before age 59 1/2)

$$E = E(prev) * (1 + f)$$

$$F = A + B - C - D + E$$

let P(prev) = principal amount at end of previous year
A = principal amount at beginning of year
B = portfolio return (amount)
C = annual income needed (amount)
C(prev) = previous year income needed
D = penalty for early withdrawal (amount)
E = annual social security and pension income (amount)
E(prev) = previous year society security and pension income
F = principal amount at end of year
i = assumed annual portfolio % return (as a decimal)
f = assumed annual % inflation rate (as a decimal)
t = estimated % tax rate (as a decimal)

15.2 The "Intrinsic Value" of the Stock Market

The indicators used in the Fuzzy Logic concept to evaluate the current equity market conditions have been described. Most investors and analysts use other methods to determine if a given stock or market is over-, under-, or fair valued. Analysts looking at the fundamentals of a company might consider among others, the expected growth in earnings, the strength of management, and the company's products' potential for the future. Although we do not use the Intrinsic Market indicator for our assessment of current stock market conditions, a short mention of this indicator is appropriate since this formula is widely used by many market watchers to determine if the stock market is fairly valued. Because of its popularity, I show the monthly updated "intrinsic" value of the Dow Jones Industrials in my home page.

let a = current Dow Jones Industrials
b = current P/E ratio of the Dow Jones Industrials
c = current yield on 10-year Treasury note
d = current earnings yield of the Dow Jones Industrials
x = % difference from fair value
Y = intrinsic value of the Dow Jones Industrials

then:

$$x = ((c - d)/d) * 100$$

(Note: a negative result = % undervalued and a positive result = % overvalued.)

$$Y = (1 - 0.01 * x) * a \text{ (use appropriate +/– sign for x)}$$

Now let's look, in hindsight, how this "indicator" worked out during the market turmoils in 1997 and 1998:

	Jan 97	*Sept 97*	*July 98*	*Oct 98*
Dow Jones Industrials	6445	7922	9318	7899
Dow Jones Industrials P/E	24.2	20.9	23.4	20.0
10-Year Treasury Yield	6.3	6.0	5.7	4.41
Intrinsic Value	3061	5910	7082	8748
% Overvalued	52.5	25.4	24.0	
% Undervalued				11.8

The first three results are history. It remains to be seen in the near future if the result of the October 1998 analysis (market 11.8% undervalued) was correct or dead wrong. Of interest: Alan Greenspan uttered the now famous words of "irrational exuberance" in the market around January 1997, i.e., when the Dow Jones Industrials stood around 6500. According to the result above (52.5% overvalued), he was not wrong in his assessment then, but the market just didn't wanted to play ball with him. As a matter of fact the market stayed overvalued for over $1^1/_2$ years before it took the big hit in August 1998 and became undervalued. What can we learn from this example? Simply this:

a) When there is too much money around, investors tend to lean toward excesses and drive the market to new highs against all rationality, irrespective of what the best brains on Wall Street have to say. The opposite can also take place, namely, "irrational consternation" could just as well set the tone, resulting in widespread pessimism and thus drive the market down to unexpected lows.

b) Any indicator might be correct from a long term viewpoint (1 to 3 years ahead) and wrong in the short term context (3 to 12 months ahead). Hence one indeed enters shallow water when attempting to predict where the market will be at a given time in the near or long term future. Alan Greenspan just made the investment community aware of the market being overvalued, he didn't say *when* the market will come down to more realistic levels. Let's give credit where credit is due.

Table 15.1 shows the relationship between a stock market's P/E ratio and prevailing interest rates, the two variables governing the so-called "intrinsic" value of the market.

A market specialist who proclaims the stock market will be at such and such a level at year's end is a fortune teller. One who tells us the market will come back quickly after a precarious drop is a dreamer, and one who tells us the market currently doesn't make any sense is a wise old man.

15.3 Reading the Annual Report

A mutual fund investor has the same privilege as a shareholder of a public company. This means he can cast his vote for a recommended slate of officers and trustees to run the fund — usually people he has never heard of and certainly doesn't know anything about their expertise and capabilities. He can also vote by proxy on proposals that would change the fund's basic investment objectives. What a shareholder cannot do is to select the fund manager nor has the shareholder a say in matters of how much salary and bonuses the manager and officers should receive. Voting by proxy is thus not a big deal to us. We usually approve all the proposals to get this obligation out of our way. More important, we vote with our pocket book. If we don't like what the fund is doing, don't like the management or its investment objectives, we just exit, sell our holdings, and move our money to another fund more representive of our own objectives.

As shareholders we also will receive the fund's annual report and this is important to us. It usually contains the nuggets we are looking for to determine if the fund meets our investment style and objectives. When the fund had a good year, the president warns us that it will not always be like that, and after a bad year, he reminds us we have to have a long term viewpoint for our investment goals.

We also don't pay too much attention to how nice the annual report looks. Glossy prints and nice pictures of the chairman walking the offices and shaking hands with subordinates don't mean much to us.

We normally skip the comments by the fund manager himself, too, because he generally will glorify the holdings that significantly contributed toward better performance of the fund and will give a ton of excuses, naturally

Table 15.1 Intrinsic Value of Dow Jones Industrials

	10-Year Treasury Bill Yield															
P/E	*3.00*	*3.25*	*3.50*	*3.75*	*4.00*	*4.25*	*4.50*	*4.75*	*5.00*	*5.25*	*5.50*	*5.75*	*6.00*	*6.25*	*6.50*	*6.75*
11.0	–67	–64	–62	–59	–56	–53	–51	–48	–45	–42	–40	–37	–34	–31	–29	–26
11.5	–66	–63	–60	–57	–54	–51	–48	–45	–43	–40	–37	–34	–31	–28	–25	–22
12.0	–64	–61	–58	–55	–52	–49	–46	–43	–40	–37	–34	–31	–28	–25	–22	–19
12.5	–63	–59	–56	–53	–50	–47	–44	–41	–38	–34	–31	–28	–25	–22	–19	–16
13.0	–61	–58	–55	–51	–48	–45	–42	–38	–35	–32	–29	–25	–22	–19	–16	–12
13.5	–60	–56	–53	–49	–46	–43	–39	–36	–33	–29	–26	–22	–19	–16	–12	–9
14.0	–58	–55	–51	–48	–44	–41	–37	–34	–30	–27	–23	–20	–16	–13	–9	–5
14.5	–57	–53	–49	–46	–42	–38	–35	–31	–28	–24	–20	–17	–13	–9	–6	–2
15.0	–55	–51	–48	–44	–40	–36	–33	–29	–25	–21	–18	–14	–10	–6	–3	1
15.5	–54	–50	–46	–42	–38	–34	–30	–26	–23	–19	–15	–11	–7	–3	1	5
16.0	–52	–48	–44	–40	–36	–32	–28	–24	–20	–16	–12	–8	–4	0	4	8
16.5	–51	–46	–42	–38	–34	–30	–26	–22	–18	–13	–9	–5	–1	3	7	11
17.0	–49	–45	–41	–36	–32	–28	–24	–19	–15	–11	–6	–2	2	6	11	15
17.5	–48	–43	–39	–34	–30	–26	–21	–17	–13	–8	–4	1	5	9	14	18
18.0	–46	–42	–37	–33	–28	–24	–19	–15	–10	–6	–1	4	8	13	17	22

Table 15.1 (continued) Intrinsic Value of Dow Jones Industrials

	10-Year Treasury Bill Yield															
P/E	*3.00*	*3.25*	*3.50*	*3.75*	*4.00*	*4.25*	*4.50*	*4.75*	*5.00*	*5.25*	*5.50*	*5.75*	*6.00*	*6.25*	*6.50*	*6.75*
18.5	–45	–40	–35	–31	–26	–21	–17	–12	–8	–3	2	6	11	16	20	25
19.0	–43	–38	–34	–29	–24	–19	–15	–10	–5	0	5	9	14	19	24	28
19.5	–42	–37	–32	–27	–22	–17	–12	–7	–2	2	7	12	17	22	27	32
20.0	–40	–35	–30	–25	–20	–15	–10	–5	0	5	10	15	20	25	30	35
20.5	–39	–33	–28	–23	–18	–13	–8	–3	3	8	13	18	23	28	33	38
21.0	–37	–32	–27	–21	–16	–11	–6	0	5	10	16	21	26	31	37	42
21.5	–36	–30	–25	–19	–14	–9	–3	2	7	13	18	24	29	34	40	45
22.0	–34	–29	–23	–18	–12	–7	–1	4	10	16	21	27	32	38	43	49
22.5	–33	–27	–21	–16	–10	–4	1	7	13	18	24	29	35	41	46	52
23.0	–31	–25	–20	–14	–8	–2	4	9	15	21	27	32	38	44	50	55
23.5	–30	–24	–18	–12	–6	0	6	12	18	23	29	35	41	47	53	59
24.0	–28	–22	–16	–10	–4	2	8	14	20	26	32	38	44	50	56	62

Note: A negative number = % undervalued
A positive number = % overvalued

all outside of his control, why some turkey investments in the funds portfolio didn't do so well.

We look in the annual report for the following information:

1. The fund's performance (total returns) in the past 1, 3, and 5 years.
2. The fund's performance as compared to the market index and to other funds with the same investment objectives.
3. The fund's top five or ten holdings.
4. The fund's asset allocation.
5. In what industry sectors (or bond sectors) is the fund most heavily concentrated?
6. Does the fund engage in hedging (options/futures)?
7. Does the fund hold foreign securities and in which countries?
8. Statement of assets and liabilities.
9. Statement of operation.

In the statement of assets and liabilities we pay attention to the size of the fund, the number of shares outstanding, and the amount of cash holdings.

In the statement of operations we focus on the management fees in particular and the expenses as a whole. The problem with most fund reports is that they are expressed in thousands (000s) and thus tend to camouflage at first glance the true size of expenses. When a report states management fees in the amount of 105,340, it doesn't appear to be too large, but when we hang on the three zeros, the amount becomes 105,340,000 and thus acquires a completely new meaning. Likewise the figure for trustees' compensation. The actual expense of 128,000 makes you think twice. Six guys meeting once a quarter or sometimes only twice year and getting paid 21333 each doesn't add up. Here the "cost-to-yield" doesn't seem to make sense.

We have modified a typical annual report and carried our calculation a little further to determine how these expenses would impact investment returns in such a fund. This is for illustration purposes only to show the reader what to look for in a typical annual report. The results of this analysis are shown in the attached Table 15.2.

In the last column we calculated our costs for expenses if we would have invested 50,000 in this fund. To have this amount invested would have cost us 373.17 in expenses for the given year. Note: we emphasize the term "our costs" because these are expenses that have reduced the return of our investment in the fund. In all fairness we have to mention the fact that the fund gave us a total return of 13.5%; its expenses were not higher than what most

Table 15.2 Fund Statement of Operation

Statement of Operations					
	NAV Beginning of Period			*19.38*	
	Number of Shares Outstanding			*1023475*	
				per share	for 50,000 invested
Investment Income	Dividends and Interest		835675	0.817	2005.40
Expenses					
Management Fees		110234		0.10771	264.38
Transfer Agent Fee		47654		0.04656	114.29
Accounting Fees and Expenses		1021		0.00100	2.45
Trustees Compensation		113		0.00011	0.27
Custodian Fees and Expenses		499		0.00049	1.20
Registration Fees		672		0.00066	1.62
Audit		222		0.00022	0.54
Legal		187		0.00018	0.44
Miscellaneous		119		0.00012	0.29
	Total Expenses	160721			
	Expense Reductions	–5126	155595	0.15203	373.17
Net Investment Income			680080	0.66400	1629.85
Net Realized Gain (LOSS) from Operations			1995776	1.95000	
	Total from Operations		2675856	2.61000	
Less Distributions			1913898	1.87000	4590.08
	NAV End of Period			20.12	
Fund Return (= the gain in NAV plus the distribution)			13.5 percent		

other funds would have shown and we should be satisfied. We are. But in the back of our mind remains the question: what if the fund has a bad year in the future, i.e., a negative return. Expenses then certainly wouldn't be reduced significantly, this we know. We are just like anybody else — we accept (with a grunt) the expenses when the fund returns are good but start to raise questions when the fund loses money. Fund expenses can also be high because the fund allows frequent switching without cost to the investor, penalizing the ones that hold their investments for the long term.

Naturally an investor will have to question if it would be better to make decisions for buying individual securities on his own. Especially if he has a personal portfolio of over 300,000. According to our example he could save himself over 2000 a year by taking this step. Why invest in a mutual fund if its performance is most likely not better than the performance of a given market index? This, too, is a legitimate question.

But investing in mutual funds has certainly a lot of positive aspects one cannot overlook. We are mutual fund investors because

- We don't have the time to research individual securities and to watch our holdings on a weekly basis.
- We are not better in picking the right securities than the fund manager himself. If we believe we can do better than the fund manager, consistently over a long period of time, we are likely deceiving ourselves.
- We don't want to pay high brokerage fees to buy and sell securities. Instead we buy no-load or low-load funds.
- We don't want to spend an inordinately long time sorting through brokerage reports to figure our taxes at year end.
- (Note: We are not adverse to consulting a financial advisor or the fund company itself to guide us in the selection of the appropriate type of funds, but the decision to buy, sell, or hold any such fund will be made by ourselves, using the investment technique described in this book.)

We have set our main attention in this book toward the small mutual fund investor. Indeed, topics discussed herein were selected for the purpose of providing the mutual fund investor with guidelines and help to control his investments. But, the book has broader applications. Investment houses, advisors, and mutual fund sales organizations could just as well benefit. We see a possibility for them to also adopt the concept of stating risk levels and representation of portfolio asset allocation for the mutual funds they manage or sell in a manner as described in this book.

We believe we have opened an opportunity in this book for mutual fund companies to show more transparency for their funds. Instead of limiting a customer to the simple explanations that this and that fund is more or less risky, they could adapt our concept and clearly state the risk level of the fund (the fund assets as a whole) in a manner as proposed in this book. A customer would then have, in the form of one number, a clear indication what risks he takes by buying a particular fund.

Gone would be the times when an investment advisor could use the standard excuse employed for poor fund performance, quote: "The fund is risky and therefore only suited for investors who can stay with it for five or ten years." As we have learned in this book, such excuses are often a ploy by fund managers and fund sales people to cover their own shortcomings, in short, using the compounding effect as if this had something to do with a fund manager's expertise. *The credit for the power of compounding and its effect on portfolio returns over a longer period of time belongs solely and alone to the investor and not the fund managers.*

To all the readers of this book we hope it will open for them new horizons in the world of mutual fund investments. We wish the herein presented guidelines will help their investment portfolio grow to a level that allows for financial security and much happiness in their retirement years.

Index

a

b

c

d

e

f

g

h

I

j

k

l

m

n

o

p

q

r

s